FEMINIST APPROACHES TO SCIENCE

THE ATHENE SERIES

The ATHENE SERIES assumes that all those who are concerned with formulating explanations of the way the world works need to know and appreciate the significance of basic feminist principles.

The growth of feminist research has challenged almost all aspects of social organization in our culture. The ATHENE SERIES focuses on the construction of knowledge and the exclusion of women from the process—both as theorists and subjects of study—and offers innovative studies that challenge established theories and research.

ON ATHENE—When Metis, goddess of wisdom who presided over all knowledge was pregnant with ATHENE, she was swallowed up by Zeus who then gave birth to ATHENE from his head. The original ATHENE is thus the parthenogenetic daughter of a strong mother and as the feminist myth goes, at the "third birth" of ATHENE she stops being Zeus' obedient mouthpiece and returns to her real source: the science and wisdom of womankind.

FEMINIST APPROACHES TO SCIENCE

edited by
Ruth Bleier
University of Wisconsin–Madison

Pergamon Press
New York · Oxford · Beijing · Frankfurt
São Paulo · Sydney · Tokyo · Toronto

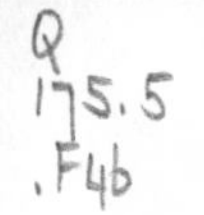

U.S.A.	Pergamon Press, Inc., Maxwell House, Fairview Park, Elmsford, New York 10523, U.S.A.
U.K.	Pergamon Press plc, Headington Hill Hall, Oxford OX3 0BW, England
PEOPLE'S REPUBLIC OF CHINA	Pergamon Press, Room 4037, Qianmen Hotel, Beijing, People's Republic of China
FEDERAL REPUBLIC OF GERMANY	Pergamon Press GmbH, Hammerweg 6, D-6242 Kronberg, Federal Republic of Germany
BRAZIL	Pergamon Editora Ltda, Rua Eça de Queiros, 346, CEP 04011, Paraiso, São Paulo, Brazil
AUSTRALIA	Pergamon Press Australia Pty. Ltd, P.O. Box 544, Potts Point, NSW 2011, Australia
JAPAN	Pergamon Press, 5th Floor, Matsuoka Central Building, 1-7-1 Nishishinjuku, Shinjuku, Tokyo 160, Japan
CANADA	Pergamon Press Canada Ltd, Suite No. 271, 253 College Street, Toronto, Ontario, Canada M5T 1R5

First printing 1986
Second printing 1988

Library of Congress Cataloging in Publication Data

Main entry under title
Feminist approaches to science.
(The Athene series)
1. Science—Social aspects. 2. Women in science.
I. Bleier, Ruth, 1923– . II. Series.
Q175.5F46 1986 591'.01 85–28378
ISBN 0–08–032787–7 Hardcover
ISBN 0–08–032786–9 Flexicover

Printed in Great Britain by A. Wheaton & Co. Ltd., Exeter

Contents

Preface

This collection of papers explores the nature of contemporary science and attempts to extend our visions toward a science that is different, better, feminist, and emancipating. Most of the papers were presented in their original form at a symposium, Feminist Perspectives on Science, at the University of Wisconsin, Madison, in April 1985.

My introductory essay attempts to highlight and weave together the main points and common themes of the nine papers, as well as point up some important contradictions.

In the first paper, Marion Namenwirth sets the stage for the subsequent papers by examining how the method and philosophy of science are supposed to work and how they do indeed work in the day-to-day realities and tensions of today's laboratories. A scientist, she suggests ways in which science could be made to work better.

Elizabeth Fee briefly traces the tradition of sexualized and genderized metaphors that have characterized science and nature and that have helped to shape views of science as male and as dominating. She describes the feminist responses to the ideologies and goals of modern science. Fee explains the importance to our theoretical and practical work of understanding other literatures of criticism and struggle against the various and interrelated forms of human domination.

Hilary Rose asks why women have been excluded from science and why science is, therefore, a peculiarly masculine institution. She grounds her answers within the theory of labor. She projects a view of a feminist epistemology, feminist approaches to knowledge of nature and people, that will be qualitatively different because it is rooted in the caring labor of women's work.

Donna Haraway examines the practices of women primatologists in the United States in the past decade and a half. She shows that the study of monkeys and apes are major areas in which feminist concerns about the relations of gender, knowledge, and power are played out. They are sciences of complex story-telling practices, through which feminist scientists contest for authoritative accounts of human origins, evolution, and behavioral biology.

Sarah Blaffer Hrdy examines classical Darwinian theories of sexual selection with their assumption of the sexually aggressive, promiscuous male and the coy, passive female. She shows the extent to which such erroneous assumptions have hampered understanding in important areas of primate social behaviors and indicates the importance that gender of the researcher has played in the cognitive history of the field of primatology.

In my paper, I use the example of research on brain hemispheric lateralization and cognitive functioning to demonstrate the damaging effects of gender-biased assumptions and interpretations in important and highly publicized areas of the neurosciences. I suggest that the language scientists use often invites the reader to supply the relevant cultural significance that the data they present would fail to support. Instead of the meaningless nature-nurture dichotomy implicit in sex differences research, I present an alternative argument for the inextricability of biological and environmental factors in the development and functioning of the brain.

Sue Rosser explores possible relationships between the phenomenal development of the field of women's studies and the marked increase in the number of women being trained in the sciences over the past decade and a half. Using a scheme tracing steps by which feminist transformations have occurred in other disciplines, she explores the changes that have been occurring in the teaching of science, in research, and in the personal development of feminist awareness among women scientists. In this she sees movement toward feminist transformations of science itself.

Mariamne Whatley emphasizes in her chapter the practical and damaging social, political, and medical consequences of biased, misogynist, stereotyped, and biologically determined thinking that pervades many of the science and health issues that capture public attention. She discusses the importance of the classroom—at the university and long before—as a place to undermine the unchallenged authority of scientific pronouncements that have consequences for human health, dignity, and potential.

Finally, Sue Searing has provided an up-to-date, but not exhaustive, list of references to feminist writings on science as a guide for further reading in the area.

The project in which we—and others not represented here—are engaged is to understand the natural world as well as the behaviors and relationships of people. But we are also preoccupied with understanding better how to gain that understanding. Readers will note that the contributors have some profound differences in approaches, attitudes, and beliefs. These relate to scientific theory and approach (for example, sociobiology) and to the nature itself of science, as a body of knowledge and as a route to knowledge. Some of these are discussed in the Introduction.

Beyond and including the differences, our routes and goals for understanding are no different than those of our feminist colleagues in other disciplines. In fact, a fair statement for most of us has been made by a biographer, Carol Ascher:

But then I imagine a new aesthetic (and a new morality) in which people, including myself, are more at ease with closeness, with uncertainty about truth, and with the confusing mix of subject and object that constitutes what is finally there to be seen—and what the reader's eye and mind take in with her or his own predispositions. I suspect that if such an aesthetic were to develop it would be accompanied by an easing of the stranglehold of fact and science, which in our day so often makes us fear a world beyond our control. (*Simone de Beauvoir: A life of freedom*, Boston: Beacon Press, 1981, p. 102)

Acknowledgements

I am grateful to the members of the Women's Studies Program – Cindy Cowden, Teri Hall, and Mariamne Whatley – who helped in the planning and organizing of the Symposium on Feminist Perspectives on Science, and to the Anonymous Fund of the University of Wisconsin, Madison, that made the symposium possible. I thank those contributors to this volume – Donna Haraway, Sarah Hrdy, Marion Namenwirth, Sue Rosser, and Mariamne Whatley – who provided me with useful criticism and commentary on the Introduction. I am also very grateful to Renate Duelli Klein and the anonymous reviewer who commented so helpfully on the Introduction and the entire manuscript.

Chapter 1

Introduction

Ruth Bleier

What is it about science—or about women—or about feminists—that explains the virtual absence of a feminist voice in the natural sciences, as an integral part *of* the sciences, with the single exception of primatology? And what would such a voice sound like? How would science be different? How would our perceptions of the natural world, of women and men, be transformed? While, over the past 10–12 years, feminists within science and without have been dissenting from and criticizing the many damaging and self-defeating features of science (the absolutism, authoritarianism, determinist thinking, cause-effect simplifications, androcentrism, ethnocentrism, pretensions to objectivity and neutrality), the elephant has not even flicked its trunk or noticeably glanced in our direction, let alone rolled over and given up.

Yet, we may ask, is that what is most important? Is our only goal to change science itself? I would say it is, indeed, a goal, but one among many. Not the least important aim, as in the rest of feminist scholarship and practices, is to win the struggle for the minds of those women, perhaps the majority, who are constrained or oppressed by internalized scientific judgments about our presumed biological limitations. For then, together we can change science, like the rest of society. The reading public, including feminists, who have been dazzled by the real and life-saving and life-improving accomplishments of science but also intimidated by its life-threatening and life-killing technological "achievements" and mystified by its esoteric language and techniques, need to gain access to its practices and familiarity with its language, methods, and theories. One route to such knowledge, short of becoming scientists, is through the writings of feminist scientists and historians and philosophers of science and other radical social critics. Their articles and books, like this one, are, as Donna Haraway writes, not simply conceived *within* a particular social and political context, they are themselves *part* of the large international social struggle about the political-symbolic structure, history, and future of Woman and women.

Then, as we work toward a better understanding of the structure and meanings of science as a practice and body of knowledge and a better vision of the kind of science we want, it is also our task to make science, and the feminist and other radical perspectives on science, accessible to new generations of students within women's studies and science classrooms, as Sue Rosser and Mariamne Whatley emphasize in their chapters.

It is within the context of that worldwide struggle against imperialist, class, racial, and gender oppressions that Elizabeth Fee would like to situate meaningfully the feminist critiques of science being raised within our Western culture. Feminists criticize the authoritarianism, detachment from nature and its "objects" of study, and the denial of responsibility for the life-threatening and nature-destroying applications of scientific knowledge and technology as stereotypically masculine. Other radical critiques of science that she cites (African, American Indian, Chinese, working class) view these characteristics of Western science as European or colonial or white or bourgeois. Part of the growing strength and effectiveness of our critiques will be, Fee believes, our ability to make links, theoretically and organizationally, with currents of opposition to racial, class, and national oppressions.

Yet it is worth noting that strong and articulate voices have long been heard within the women's movements and feminist scholarship speaking out on the issues of multiple dominations, paying attention not only to gender but to race, ethnicity, class, and sexuality (in particular, heterosexism). In contrast, with rare exceptions, nonfeminist Marxist and nationalist radical thinkers and leaders, sensitive to issues of class or race or nationality, consider women themselves, the "woman question," or problems of gender quite irrelevant if not antithetical to their immediate goals of "human" liberation. However radical with respect to issues of race, ethnicity, or class, still invisible to them is their own patriarchal privilege and power, which they will not surrender meekly and voluntarily, any more than will the class of dominant white males who hold power in this country. "Joining with" historically (to the present) has meant "being subsumed within" male revolutionary movements. Without an independent voice, independent agendas for action, independent analyses of oppression that make gender no less important or urgent than any other oppression (or that even put gender first since no one else does), feminists will lose the battle once again.

The papers in this book start from certain assumptions. Feminist analyses of important areas of research in the natural sciences (both contemporary and historical) have revealed many levels of distortion and bias in assumptions, methods, and interpretations. Yet, on further analysis, this discovery need not be surprising. Science is a socially produced body of knowledge and a cultural institution. Our culture is deeply and fundamentally structured socially, politically, ideologically, and conceptually by gender as well as by race, class, and sexuality. It then follows that the dominant categories of cultural experience (white, male, middle/upper class, and heterosexual) will be reflected within the cultural institution of science itself: in its structure, theories, concepts, values, ideologies, and practices. As Fee writes, science is done by means of human relationships, and these relationships are unequal in terms of power across several different boundaries—of class and race as well as of gender. "By necessity, all of these power relations are reproduced within scientific knowledge; the scientist, the creator of knowledge, cannot step outside his or her social persona. . . . Reflected within science is the particular moment of struggle of social classes, races and genders found in the real, natural, and human world" (Chapter 3). Each scientist has a particular history of experiences and social relationships and, therefore, a particular worldview and set of values, beliefs, hopes, and needs that are reflected—as they are for everyone else—in what scientists do, how they (we) perceive the world, how they view and experience social relationships and questions of power, and how they practice their science.

There is an oral tradition and set of idealized practices known as the scientific method, which includes making observations, forming hypotheses or tentative explanations for the observations, and then testing the validity of the hypotheses by further observations or experiments. This method is generally viewed as the protector against rampant subjectivities and the guarantor of the objectivity and validity of scientific knowledge. Yet each step in the scientific method is profoundly affected by the values, opinions, biases, beliefs, and interests of the scientist. These values and beliefs affect what observations scientists make and, therefore, what they believe needs explaining in the world, what questions they ask: for example, what are the origins of gender differences in status and roles or the origins of IQ differences between blacks and whites? They affect the assumptions scientists make: what language they use to pose questions; what they see and fail to see; how they interpret their data; what they hope, want, need, and believe to be true.

While we all have the experience of not seeing things that are before our eyes, it still may be difficult for most nonscientists (or scientists) to believe that scientists' values, beliefs, and expectations can influence what they are actually able to *see* or *hear* with their perfectly functioning senses. For example, leading microscopists of the 17th and 18th centuries, including the great van Leeuwenhoek, claimed they had seen "exceedingly minute forms of men with arms, heads and legs complete inside sperm" under the microscope. Their observations were constrained not by the limited resolving power of the microscopes of the time, but rather by the 2,000-year-old concept, dating from the time of Aristotle, that women, as totally passive beings, contribute nothing to conception but the womb as incubator. Except for Japanese field workers, primatologists in the 1950s and 1960s could not *see* what female primates were doing; and even if they could see something, their hypotheses, observations, and interpretations were clearly constrained by the cultural concepts available. Attempts to explain female leadership, or dominance, or sexual aggressivity and initiative had to be accommodated within male-centered explanatory systems.

As human beings we *all* have deep-seated beliefs on most important issues and those beliefs derive from sources that are other than scientific. In turn, when such issues, like gender or race differences, become the subject of scientific investigation, scientists are not magically capable of suspending belief and judgment in their approach to the problem. And this is true for scientists on all sides of any controversy. Suspensions of belief, emotion, and judgment are no more possible in the pursuit of answers about gender than they were in the race to reveal the structure of DNA (or, more accurately, the race to win the Nobel Prize) by Watson, Crick, and Wilkins. There is hardly a significant area of science, however remote from gender or race or other social issues, that does not engender wildly differing opinions, intense passions, irrational responses, and personal antagonisms. And, furthermore, I would maintain that most scientists would not be happy in their work nor would science have accomplished so much as it has in understanding natural phenomena and applying the knowledge to social uses were these passions and drives absent from the laboratory.

The problem is that, more often than not, these passions and commitments have more to do with drives for personal power than with the pursuit of the truths of nature. And when the questions being investigated have important social implications about the "nature" of women, the commitment is to the social status quo rather than to a disinterested and unemotional consideration of the range of possible interpretations of a body of observations. That is, scientists, like the majority of men in our society, have a personal stake in a system and ideology that reinforce belief in the biological inferiority of women and, thus, justify women's subordinate position within the home *and* the laboratory. (The phenomenon of the eminent scientist could hardly exist without a veritable army of unpaid and underpaid women who have facilitated his career—as wife, technician, secretary, and underacknowledged scientific colleague contributing essentially to his research.) Yet scientists would still maintain that what they do in their laboratory is neutral, objective, and value-free; and that their differences of opinion, emotions, and drives are objective—based on rational differences in techniques or interpretations—all quite separate from who they are as people. In her chapter, Marion Namenwirth provides a glimpse of the biological sciences as practiced in the contemporary United States, with insights into the everyday workings and pressures of the modern laboratory.

The field of the sociology of scientific knowledge attempts to relate the production of scientific knowledge by scientists to scientists' social, political, and personal contexts. Sociologists of scientific knowledge (Karin Knorr-Cetina, Michael Mulkay, G. Nigel Gilbert, Bruno Latour, and others), through their participatory observations of daily life in particular science laboratories, record and analyze the patterns of daily scientific laboratory work and conversations, written communications with other scientists, lectures, and publications. They have documented the commonplace fact that scientists—in their drives, motives, and operations—are influenced by many of the same considerations that affect others trying to get ahead and keep ahead in the world—competitiveness, envy, and dependence on recognition and rewards. The subfield of linguistic and discourse analysis seeks, in the words of its chief practitioners, Gilbert and Mulkay, to answer the question, "How are scientists' accounts of action and belief socially generated?" (1984, p. 14). They describe scientists' recurrent interpretative practices—for example, successive descriptions (in letters, lectures, and publications) of a particular experiment or of a particular "finding"—and show how these descriptions change in accordance with successive variations in their social context. An example of this phenomenon in the area of hormonal effects on the developing brain is described in my chapter in this volume.

(It is noteworthy, incidentally, that the sociology of scientific knowledge, with the exception of the work of Sharon Traweek cited in the chapters by Fee and Haraway, remains peculiarly oblivious, so far as I can tell, to the entire body of feminist scholarship in all disciplines, which contributes fundamentally, both theoretically and methodologically, to the sociology of knowledge including scientific knowledge; nor does it appear to recognize gender as a sociological category of possible significance in understanding the social production of scientific knowledge.)

Yet, in the face of all of this, we know that the objectivity and social neutrality of science and scientists are supposed to be what distinguishes the pursuit of knowledge by scientific means from that pursuit by other means. Neutrality is believed to be an inherent and defining feature of science. It is an interesting paradox, however, that even the idea itself of objectivity and social neutrality as a characteristic or even requirement of science is not an inherently logical, internal achievement of modern science, but rather the product of social and political forces in the 17th century, in the view of van den Daele (1977). A part of the reform and New Learning movements of Puritan England prior to 1660, Baconian science represented one aspect of the battle against the philosophical authority of the ancients, against received wisdom. Its antielitism and antiauthoritarianism were expressed in its emphasis on empiricism, on experimentalism (manual labor integrated with intellectual labor) as essential features of natural philosophy. Furthermore, Baconian science was associated with the goals of social, political, and educational reform and emancipation. But the end of the Puritan Revolution, with the Restoration in 1660, terminated this association of science with social and political concerns. This separation "is a historical compromise that follows not only from scientific purposes but also from the exploitation of opportunities for institutionalization offered within the context of Absolutism" (van den Daele, 1977, p. 45). The separation of science from social reform was essential for retaining royal protection and support for the important processes of the institutionalization of science, begun with the establishment of the Royal Society in London in 1662.

Retained from Baconian science was the emphasis on experimentalism, which could be defended as culturally neutral because of its universalistic character, and its association with utilitarian concerns, which were important to Absolutist rulers. Thus, in van den Daele's view, "The normative (social, political, religious) neutralization of the knowledge of nature, which for us is an essential element of the 'positive,' objective, and concrete character of scientific knowledge, was a condition for the institutionalization of science in the seventeenth century" (p. 28). His thesis suggests "that there is a connection between the rise of science as a cognitive program and the rise of science as a social structure" (pp. 28–29). Thus, even though there exists for us a categorical distinction between social reform and knowledge of nature, 17th century science, in fact, *made a choice* for the institutionalization of a positivist science, objective and neutral, to the exclusion of alternative cognitive programs of knowledge of nature as well as alternative social structures of science, such as those that "would at least institutionally and normatively be associated with the requirements of human progress, enlightenment and emancipation" (p. 47).

It is striking and ironic that while the enlightened and emancipatory goals of 17th century Baconian science were aborted, Bacon's influence on molding science in the image of men and masculinity was considerable and long-lived—with us, in fact, to this day. Though Bacon rejected all *other* kinds of recognizable established authority, he accepted and *established* male authority as integral to the practice and philosophy of science. Continuing a process begun at least in the 16th century (Fee, Chapter 3), Bacon elaborated the metaphors of science in sexual and gendered terms, with science as male and nature as female, a mystery to be unveiled and penetrated. Woman as a reproductive being embodied the natural, the disordered, the emotional, the irrational; man as a thinker epitomized objectivity, rationality, culture, and control. The subject of these dualisms and metaphors has been explored by feminist philosophers and historians of science, as Fee recounts, and figures in several essays in this book because of their importance in the molding of gender-differentiated stereotypes and in the structuring of science as male, both conceptually and organizationally. Since ideas are not generated in cultural vacuums, the exclusion of women from the practice of science and the consequent male, patriarchal structuring of science is reflected in the concerns, concepts, metaphors, assumptions, and language of science, a claim demonstrated in a variety of ways in most papers in this volume.

There could be no finer demonstration of the centrality of gendered metaphors in science than the passage quoted by Fee from Richard Feynman's speech on receiving the Nobel Prize. That was in 1966, but misogynist arrogance has continued to thrive among Nobel laureates in science. James Watson was the wonderboy from Harvard when, with Crick and Wilkins, he won the Nobel prize in 1962 for describing the double helical structure of DNA (this ambitious triumvirate "received" their insights into—penetrated—DNA's structure soon after illicitly and secretly viewing Rosalind Franklin's unpublished crystallographic images of DNA). Watson recently bubbled over to reporters in a fit of petulance over the federal regulation of genetic engineering:

> One might have hoped that the Republicans would have been more sensible about regulations, but they were just as silly as the others. . . . The reason is that the White House receives its advice from people who know something about physics or chemistry. The person in charge of biology is either a woman or unimportant. They had to put a woman some place. They only had three or four opportunities, so they got someone in here. It's lunacy. (quoted in *Science*, 12 April 1985, p. 160)

Namenwirth points out that because the science that we know evolved within a patriarchal society, it developed a decidedly masculine tone, became distorted by a pervasive male bias, systematically excluded women from training and participating in science (as in all other professional and public activities), and was, furthermore, most effective in propagating stereotypes of "the feminine" that made it seem self-evident that women were totally unsuited for "penetrating" nature's mysteries. Despite that mighty effort, substantial numbers of women became scientists in the latter part of the 19th and early 20th centuries, always, at first, under the sponsorship and protection of a father, brother, or husband who was a scientist, as Hilary Rose describes. Later communities of affluent women lent their money, names, and aggressive talents to opening up possibilities for the training and employment of women as scientists. As Margaret Rossiter (1982) documents with many painful details, most of these women, however accomplished and productive, struggled, often throughout their lives, for places to work, salaries, and acceptance by their profession. Some achieved recognition, even Nobel prizes, and some saw their work, as happens to this day, appropriated and subsumed by laboratory chiefs, whose interest in cheap labor was entirely and rewardingly compatible with their need for innovative ideas and accomplished scientific productions, whatever the source.

Hilary Rose analyzes the exclusion of women from science (that is, as equally recognized practitioners; the majority of people actually *practicing* science, technicians, *are* women) as a particular instance of the general division of productive and reproductive labor between men and women, a division that is part of the general division of knowledge between men and women and is reflected in differences in paid and unpaid labor and in amount of labor time. With women concentrated in low-paid work within science, as they are in the rest of a segregated labor market, the research faculty in both the United States and Britain is overwhelmingly male, white, and middle class. This predominance of males in science research "has resulted in excessive reliance on conceptual paradigms related to social preoccupations of this class of males" (Namenwirth, Chapter 2).

Nowhere have those preoccupations been played out with more drama and intellectual and social consequence than in the field of primatology, as both Donna Haraway and Sarah Blaffer Hrdy describe in this volume. One could say that any contemporary body of knowledge, for example, the literary or musical canon, molecular biology, econometrics, sexology, labor history, is not necessarily the most accurate reflection of the reality it purports to describe, but instead the (temporarily) victorious product of struggle among contestants for authoritative control in the field. That this is "politics by other means" could not be made more clear than by an analysis of the epistemology and contemporary sociology of the field of primate social behavior, as Haraway presents in her chapter. Through the use of "narrative strategies and the social power to deploy them for particular audiences," primatology presents "the life stories of monkeys and apes [as] industrial and postindustrial versions of the past and the future of them and us." It thus participates "in the preeminent political act in Western history: the construction of Man." That is, far from being simply a descriptive science based on the observed and interpreted behaviors of primates, primatology is a powerful narrator of authoritative origin myths. In Haraway's words, it is a political discourse about the foundations and origins of sociality, a major social practice to construct and negotiate boundaries between self and other, human and animal, gender and sex, culture and nature, men and women. Thus, it is a science of difference, just as politics is about difference, gender is about difference, and the "West" is about difference.

Beginning in the 1970s, the entrance of women and feminists into primatology as the observers who reconstructed the meanings of female primates disrupted a previous whole, disrupted a previous version of "we," the white Western brotherhood, embodied in the male primatologist. The *we* became all female primates, observed and observers, participating in the construction of a new definition of *female* and, therefore, *woman* and, in Haraway's words, also raising the cost (for example, requiring better field data) for making claims about female biology and behavior. In the process, the female primatologist (hence, woman) was reconstructed to be a citizen, one who constructs knowledge, a scientist.

Hrdy uses the example of sexual selection theory and primatologists' uncritical adoption of its culturally stereotyped formulations of the active, courting, promiscuous male and the passive, coy, faithful female to demonstrate the mind-numbing and eye-blinding effect that a ruling paradigm can have in a field. Even when primatologists looked at female primates, they could not see the full extent of female sexual choice, initiative, and aggressivity or its polyandrous expression because they had no conceptual framework or language through which to describe or interpret the phenomena. Behaviors become aberrations to be ignored if they seem inappropriate for existing theoretical schemes. Nothing was written about the sexual agency of female primates until 1979 (Haraway: "female primates got orgasms in the 1970s because they needed them for a larger political struggle"), and the new visibility of females requires a serious rethinking of the bases of breeding systems among primates. Similarly, the widespread involvement of males with infant care, long invisible, is only now being studied and calls into question sociobiological theories about universal differential "investment" in offspring by male and female primates.

What is important to conclude from this newer female-oriented, female-produced body of knowledge is not simply that female primates can be as dominating, aggressive, competitive, and mobile as males. While these observations are indeed important for uncovering behaviors heretofore invisible, they are still judgments that carry with them their own set of ethnocentric values and subjectivities. The observations are fitted into the language and the restrictive interpretive framework of E.O. Wilsonian sociobiology, a biologically determinist, quintessentially Western industrial science that uses concepts of genetic "strategies" and "investments" and survival of the fittest. Furthermore, primatology is the one life science, Haraway noted in a symposium panel discussion, that has regularly reconstituted heterosexuality and the monogamous nuclear family as norms for primate behavior, ignoring much available evidence to the contrary. In regard to heterosexist bias, however, primatology does not stand alone, since all modern dominant theories of human evolution also assume, without the slightest hint of evidence, that we have evolved culturally not only *from* but *because of* the organization of our australopithecine ancestors of 3–5 million years ago into nuclear families, characterized by the very familiar modern stereotype of the patriarch bringing meat to his own dependent and sedentary nursing woman. No less tenuous have been the efforts of some reproductive endocrinologists to produce hormonal models of homosexuality in rats. The uncritical and simplistic motivations for the research are the belief that homosexuality reflects a hormonal pathology and that what hormonally manipulated rats do in a cage has relevance to human sociosexual complexity.

Of significance in the newer feminist scholarship in primatology is that the previously missing viewpoint produced a body of knowledge that overturned long-held basic tenets and unquestioned assumptions in the field that formed the platform for the formulation of the most influential male-centered theories of human cultural evolution. The example of field primatology also tells us something about the sources of scientific knowledge, something about "seeing" as it relates to one's location within time and culture. In the absence of knowledge about female primates based on observations of their behaviors, primatologists then felt free to speculate about (that is, to construct) female primates in ways that allowed their imagined behaviors and characteristics to fit existing male-centered theories of human cultural evolution and thus to embellish, naturalize, and reinforce the social construction of human female and male genders and of relations of domination and subordination.

Primatology thus serves as an example of the corrections that a feminist perspective can effect in a field of knowledge. A field comparable to (pre-feminist) primatology, dominated and constrained by a ruling belief system, is the area of the neurosciences that searches for differences in the structure and function of the brain as explanations for presumed sex differences in cognitive functions and other gender associated characteristics, a subject I explore in this volume. While ideas about gender differences in characteristics, abilities, temperament, tasks, and responsibilities had long been concerns of Western philosophies, they were not explicitly formulated as part of scientific theory until the mid-19th century (Fee, 1983). Previously, Fee writes, the ruling classes depended on religious authority to legitimize social hierarchy and maintain social order. As both Elizabeth Fee (1979) and Stephen Jay Gould (1981) documented, 19th century brain scientists, neuroanatomists and craniologists, became obsessed with the measurement and remeasurement of the human brain and cranium to prove the biological inferiority and apelike characteristics of the female and "negro" brain and temperament.

As I describe in my paper, today's theories and studies are no less suspect, no less the product of research distorted by social values, hopes, and expectations. The dominant theory assumes that significant cognitive differences exist that can be explained by sex differences in the development, structure, and functioning of the brain. Studies of prenatal hormonal effects on the developing brain and of hemispheric lateralization of cognitive functions are widely accepted as providing evidence in support of the theory. Yet in such studies, speculations are rarely subjected to serious testing, but they are given credence far beyond the quality and quantity of supporting evidence. Contradictory data are ignored and alternative interpretations are not suggested. By applying insights of literary theorists on textual criticism and the role of the reader in investing the text with meaning, I suggest that the scientists whose work I quote stop just short of making assertions that their data cannot defensibly support. But they can rely upon their readers—other scientists, science writers, and the science-reading public—to supply the (intended) relevant cultural meaning to their text; for example, that women are innately inferior in visuospatial and mathematical skills.

This commitment to a gender ideology and to gender differences in scientific research has great force and implacability for several reasons. It is a scientific commitment identical with a personal, individual commitment of some scientists and a collective social ideological commitment; that is, it has great public sanction as a subject of investigation that *everyone* understands and most find comfort in; it is money well spent. As I discussed previously, that commitment is to the gendered status quo in society at large and within university departments and laboratories in particular. I would contend that there is subtle (and often not so subtle) interplay between biological theories that imply the natural inferiority of women and the conditions under which scientists and women live and work. A clear line cannot always be traced (though frequently it is all too apparent) between how scientists feel, believe, and behave toward women and the scientific theories they postulate concerning human behaviors and women's "nature." And I do not believe that male scholars and scientists have had to conspire together to distort the truth or to create theories that will protect the privileged position men enjoy in this society vis-à-vis women of their class. There are, however, necessary connections between subjective belief or experience and theories, whether formulated by Supreme Court justices or scientists. They share a world view formed in a society that takes as natural the dominance of men and the subordinance of women. It affects the ways in which physicians may view women patients and women physicians, the interactions between women and men in science laboratories, and the attitudes that male science faculties have toward the admission of women as colleagues. Biological theories about women are a necessary part of the system and set of conventions by which all-male science faculties justify their continuing exclusion of women as colleagues. (Women are being admitted in ever increasing numbers as graduate students in science as elsewhere because they are among the superior students; at the same time, women continue to constitute a miniscule proportion of science faculties.) How, then, can it be otherwise that many male scientists (and, unfortunately, some women scientists) will find a fascinating question to be, Why can only men (and a few exceptional women) practice science well? Thus, there is a social-political context within which particular scientific or medical views of women will flourish. At the same time, these views become part of that context and contribute to an ideology that permits and encourages practices that oppress women in our society.

But, then, one might be tempted to ask, if there are indeed no known biological differences between women and men in cognitive and other abilities and potentialities, would science, transfused with women and feminists, be any different and, if so, how? Does substituting a female bias for a male one make science better or only different? The papers here provide a variety of answers to these questions, and I shall start with the first question.

Sue Rosser supplies us with some optimism as she surveys the developments in women's studies and science since the early 1970s. She provides insights into the processes by which feminist transformations have occured, and probably will continue to occur within science, as she makes connections between a number of factors: the growing numbers of women entering science; the development of their awareness of the silenced voices of their scientific foremothers, of the inequities in training and employment opportunities, and of gender ideologies implicit in scientific theories and practices; the effects of these awarenesses on their own research and teaching; the finding of other questioning women in science and women's studies; the discovery of themselves and their thinking as outside existing paradigms and their consequent testing of those paradigms; and, finally, as in primatology, the transformation of the paradigms.

Mariamne Whatley implicitly believes that the feminist critiques of science and biological determinism and the visions for a feminist science can have transformative effects on science if they are made accessible to the reading and science-consuming public and, in particular, to the coming generations of scientists. Having more women scientists is not enough if they have been trained to think about science and practice science within the same authoritarian, deterministic framework that prevails today. She sees the educator's task as undermining at every level of classroom teaching (college and high school may be too late to get women into science!) the unchallenged authority of science and experimentalism, the unquestioning acceptance of the results of any scientific research. Whatley stresses the importance for the scientist-teacher of creating distrust of simplistic biological explanations of complex phenomena and helping to develop more complete and complex hypotheses. Teachers and students can together explore the question of *why* biological explanations for human behavior are dominant today: whom do they benefit, whom do they disadvantage and penalize? She points out that even "good" science, that is, research done with the highest standards of scientific methodology, reflects the values of the scientist. Thus, she challenges the notion that certain areas of research, for example, sex differences (or race differences) in cognitive abilities, can ever be done *well* and, therefore, ever be valid as a "problem" to be investigated, at least in cultures such as ours. Indeed, an important question to ask, also raised in my paper, is why such an enormous amount of energy and money is spent on the issue of sex differences in cognitive abilities when review articles and the best of experiments show either no differences or differences that are trivial in comparison with those *among* girls, *among* boys, *among* men, and *among* women. If scientists are interested either in factors affecting cognitive functioning or in improving the quality of life, surely more significant information could be gained by exploring the complex of social, economic, and political forces that account for the enormous range of abilities *within* groups.

In raising this issue, Whatley implies that feminist perspectives and goals in science would tend to eliminate from research consideration those questions that cannot possibly be answered by biological explanations in isolation from sociocultural factors (which cannot be measured or controlled) and that, by being raised, help to create an entity (biological gender or race differences in ability) that does not exist.

Furthermore, as Namenwirth asks, when there is such a multitude of interesting problems to investigate, the solution to many of which would benefit various segments of our society, how do we justify working on research whose applications have been and threaten to be profoundly destructive of natural resources, human life, and the dignity and self-respect of racial, ethnic, or gender groups? But, clearly, if feminists – along with ecologists, antinuclear, and antiracist activists – make explicit our social values in our research choices and encourage others to do so, we will be accused of attempting to impose limits on the democratic right to freedom of choice. Yet, when 72% of all federal funds for research and development (in 1986) will go into defense (*Science*, 9 April 1985); when billions of research dollars go to the space program; when many levels of professional, economic, and political (that is, nonscience) imperatives influence or determine a scientist's priorities, freedom of choice does not seem to be a robust entity in danger of sacrifice. In brief, the broad decisions about what research will be funded are made within the halls of social and political power.

Namenwirth anticipates that pressures for change in scientific practices and thinking will come, for the time being, from outside science, from the feminist movement, rather than from within, even from women scientists whom she does not see as alienated from the thinking or authoritarian voice that characterize contemporary science. But eventually, she believes, feminist and other radical attitudes about ways to do science (like some she proposes) will come to seem less strange and threatening to scientists.

Rose and Hrdy, however, believe that the presence of more women and more feminists in the sciences will itself provide the force for change, at least to the degree that it has in primatology, and perhaps even further. Rose is insistent on the value of experiential knowledge, on the basis of knowing as being in labor, and on the particularity of women's knowledge based historically in caring labor, in people work, however alienated, boring, or menial it may be. Thus, she says, "As a profoundly sensuous activity, women's labour constitutes a material reality which structures a distinctive understanding of the social and natural worlds." For her, a feminist epistemology must, therefore, represent a truer knowledge, transcending dichotomies, insisting on the validity of the subjective, and preferring complexity rather than reductionism and linearity as explanatory frameworks. If women constituted "nothing less than half the labs," the daily conversation and the work would necessarily be different, and a significant part of the battle will have been won.

Both Hrdy and Haraway agree that the entrance of large numbers of women into field primatology and their individual and collective feminist consciousness were responsible for upsetting long-held beliefs and assumptions (for Hrdy, paradigms; for Haraway, narratives and myths of origin). Women primatologists, Hrdy recounts, using herself as an example, identified with the female primates they were observing and with their problems, at the same time (in the 1970s) that they began to be aware of and to articulate problems that women confront in their world. Consequently, they began to formulate questions that had never been asked before concerning the behaviors and coping mechanisms of female primates. These new questions and the observations and interpretations made possible by them transformed a body of beliefs that had been central to primatology. Thus Hrdy would say, I believe, that primatology is now a better, a more inclusive science than it was 15 years ago. At the same time, recognizing that all people must bring some set of values and biases to their research, she anticipates that cultural diversity among researchers, multiple studies, and restudies will ensure the self-correcting processes that she believes have always characterized science. Trained within and herself using sociobiological explanations, as discussed in a symposium panel session, Hrdy recognizes the limitations that the language and assumptions of sociobiological paradigms may place upon the range of possible interpretations of animal behaviors. It is an example of the self-correcting processes she relies upon that she views the new observations and viewpoints introduced by feminist primatologists as opportunities to ask new and basic questions and reformulate sociobiological concepts. I must note again, however, that primatology is a lone example in the natural sciences of dramatic changes under the influence of feminist viewpoints. This is related, in part, to the presence of a critical mass of women and feminists within the field, a situation that does not exist in any other area of the natural sciences. Further, in general, there does not really exist a free marketplace of competing ideas, where the best ideas necessarily win out or where serious criticism of an area of deeply flawed and value-laden research is welcomed, let alone published. One example would be my own unsuccessful efforts to have the criticisms appearing in Chapter 7 of this volume published in a leading journal (*Science*) that published some of the most influential, and seriously flawed, studies on sex differences in cognitive functioning.

While Haraway would, I believe, agree that primatology is a better science for the perspectives, and story-telling, introduced by feminist investigators, she rejects ideas of "truer" successor sciences replacing the old ones or of unities and harmonies among the diverse contemporary discourses. The value of our intellectual probings and criticisms of contemporary society lies in being able to see and hear, not to reconcile (impossible in any case), the simultaneously true and contradictory multiple realities in a worldwide system undergoing profound crises: realities of capitalism, race, gender, nationalism, multinationalism, etc. What exists in scientific fields, like primatology, are tensions and contestations for authoritative power and audience among different and contradictory narratives, stories of origin, myths. We, feminists of science, are engaged in political contests for meaning, which will work not by replacing one paradigm with another, but by altering the narrative field—a totally different process. "Redistributing the narrative field by telling another version of a crucial myth is a major process in crafting new meanings." The radical project in which feminist critics are engaged is constructing "a different set of boundaries and possibilities for what can count as knowledge for everyone within specific historical circumstances. . . . Feminist science is not biased science, nor is it disinterested in accurate description and powerful theory. My thesis is that feminist science is about changing possibilities, not about having a special route to the truth about what it means to be human—or animal" (Haraway, Chapter 5).

There are theoretical dissonances among the papers in this book. Not the least basic of the differences has to do with this question of the nature of science itself and the routes to scientific knowledge. Perhaps alone among the contributors, Hrdy and Namenwirth believe in, have faith in, *science* as "a system of procedures for gathering, verifying, and systematizing information about reality" (Namenwirth, Chapter 2) and in *the scientific method* as inherently good and true. Science and its methods may come to be abused and misused by persons and groups in positions of power and authority and, thus, produce knowledge that is incomplete and biased and also deteriorate human life, nature, and dignity. Namenwirth states that science is not "inherently masculine." (I take *masculine* here to refer to the set of socially constructed characteristics attributed to men in the patriarchal cultures that are the context for our analyses.) Yet both authors describe powerful examples of mainstream dominant scientific theories, approaches, methods, "facts," and interpretations that are inescapably masculine (androcentric, patriarchal). How, then, do they escape being inherent to science? If science as a method and body of knowledge is, as it must be, a cultural and social product, how could it, unlike all other cultural products, escape the culture's most basic gendered concepts, woven into its very fabric however invisible they may still be to our own culture-bound minds? That is, how does one distinguish masculine products of science (for example, "male-authority paradigms") from a science that is "not inherently masculine"? More importantly, and intimately related to the problems facing feminist scholars in all fields, how can we even begin to conceptualize science as nonmasculine, as somehow transcendentally pure and objective (nongendered), when most of written civilization—our history, language, conceptual frameworks, literature—has been generated by men? Who is the authority that, standing above the fray, has guaranteed that science alone is untainted by androcentric biases and patriarchal concepts and methods?

It may be valuable for me at this point, before concluding, to attempt a summary of some principles that might characterize feminist science. I think the contributors to this volume would agree on them—many appear in one or more of the papers—at the same time that we would view them as inadequate, because we know we have just begun to see what we do not yet know.

We would first of all insist that scientists acknowledge that they, like everyone else, have values and beliefs, and that these will affect how they practice their science. The next task is to convince them to explore and understand in what ways these subjectivities specifically affect their perspectives and approaches, their actual scientific methods. Scientists would have to be explicit about their assumptions; honest, thoughtful, and careful in their methods; open in their interpretations of each study and its significance; clear in describing the possible pitfalls in the work and their conclusions about it; and responsible in the language used to convey their results to the scientific and nonscientific public.

Then we, as feminist scientists, in making explicit our own social values and beliefs where they are relevant to the science we practice, may wish to claim a feminist approach to scientific knowledge that in its language, methods, interpretations, and goals, acknowledges its commitments to particular human values and to the solution of particular human problems. This would not eliminate or censor basic scientific investigations done for the sake of knowledge itself, with no known practical, social application, but it *would* aim to eliminate research that leads to the exploitation and destruction of nature, the destruction of the human race and other species, and that justifies the oppression of people because of race, gender, class, sexuality, or nationality.

Our science would not be elite and authoritarian and, therefore, it would have to be accessible—physically and intellectually—to anyone interested. It would be humble and acknowledge that each new "truth" is partial; that is, incomplete as well as culture-bound. Recognizing that different people have different experiences, cultures, and identifications (therefore, different perspectives, values, goals, and viewpoints), feminist science would aim for cultural diversity among its participants, so that through our diverse approaches we would light different facets of the realities we attempt to understand. Such diversity would help to ensure sensitivity of the scientific community to the range of consequences of its work and thus its responsibility for the goals of science and the applications and by-products of its research.

For many of these changes to occur, scientists would have to learn to reconceptualize science, its methods, theories, and goals, without the language and metaphors of control and domination. And for *this* to happen to a significant degree, profound changes must occur in a system that is based on power, control, and domination and that recognizes and rewards those who support and reinforce its ideologies and aims. However, just as destroying all nuclear weapons and outlawing their manufacture is not equivalent to world peace, but nonetheless an urgent first step, so can feminists continue to take our first steps toward the transformation of the ideologies and practices of the modern institution of science.

Feminist science, being a better science, recognizes the true complexity of nature and of each individual human nature. It constantly resists efforts to reduce explanations of complex phenomena to single causes and to strip human behaviors and characteristics of the social and political contexts within which and from which they developed.

Now, one might say that what I have described is simply good science, not just feminist science. That is true, in the same way that feminist scholarship in all fields has made them *better*, opened them to new perspectives and to previously ignored experience, and radically introduced gender as an unavoidable category of analysis. Whether feminism can bring other profound transformations in what we call science is a question to be answered over the next few years.

Women in science, in the United States, and throughout the world, are empowering ourselves to question and to change the historically developed but truly bizarre circumstance that a tiny fraction of the world's people—white, Western men—can set standards of behaviors and attributes for the rest of humanity. We have accepted their definitions of science and their goals for science, but together we all pay the consequences.

The writers here present different, and still far from complete, alternative feminist visions of science. What is common is the commitment to feminism both intellectually and politically, to theorizing and to practices that can help change the lives of women in general. Each of us constantly shifts our grounds for operations within the broad boundaries of the trajectory of our life experiences. We choose the means of acting and speaking with which we are comfortable and effective, and we have viewpoints that grow out of our specific complex of cultural history. From these papers and the symposium's panel discussions, it is clear that there will be constant shiftings and shadings of viewpoints and meanings and redefinitions of directions and goals as feminists re-create a science that will only benefit rather than oppress.

REFERENCES

Fee, E. (1979). Nineteenth century craniology: The study of the female skull. *Bulletin of the History of Medicine, 53*, 415–433.

Fee, E. (1983). Women's nature and scientific objectivity. In M. Lowe & R. Hubbard (Eds.), *Woman's nature: Rationalizations of inequality*. Elmsford, NY: Pergamon.

Gilbert, G. N., & Mulkay, M. (1984). *Opening Pandora's box*. Cambridge: Cambridge University Press.

Gould, S. J. (1981). *The mismeasure of man*. New York: Norton.

Rossiter, M. W. (1982). *Women scientists in America: Struggles and strategies to 1940*. Baltimore, MD: The Johns Hopkins University Press.

van den Daele, W. (1977). The social construction of science: Institutionalisation and definition of positive science in the latter half of the seventeenth century. In E. Mendelsohn, P. Weingart, & R. Whitley (Eds.), *The social production of scientific knowledge*. Dordrecht, the Netherlands: D. Reidel.

Chapter 2

Science Seen Through a Feminist Prism

Marion Namenwirth

> Feminism means finally that we renounce our obedience to the fathers and recognize that the world they have described is not the whole world. Masculine ideologies are the creation of masculine subjectivity; they are neither objective, nor value-free, nor inclusively "human." Feminism implies that we recognize fully the inadequacy for us, the distortion, of male-created ideologies, and that we proceed to think, and act, out of that recognition. (Adrienne Rich, 1977, p. xvii)

PROLOGUE

In the course of my apprenticeship, research, and teaching academic within biology, I've come to know scientists who are honest, thoughtful, and independent, as well as those on the make. I've encountered research programs that are creative, finely crafted, and carefully executed, as well as those that are imitative and carelessly done. Yet, despite its diversity, the academic science community has a structure and themes that repeat themselves. Though it has its mavericks, science also has its usual ways of doing business. I'm fascinated by science and deeply admire certain of its practitioners. Yet, I find many aspects of the contemporary science system repugnant, anticreative, and threatening to human life and freedom. In the essay that follows, I delineate what disappoints me about academic science and sketch how I think feminism might bring about improvements for women and men alike.

Many of the ideas expressed in this paper are controversial. Note also that these ideas are based primarily on my familiarity with research in the biological sciences. I hope that those readers who are scientists or who have a keen interest in science will be stimulated to make a personal assessment of whether the science we practice today has not strayed unacceptably far from the science of which we would like to be part.

WHO ARE SCIENTISTS—AND WHY ONLY THEY?

Science is a system of procedures for gathering, verifying, and systematizing information about reality. The knowledge that has been developed in fields such as physics, astronomy, biology through scientific procedures is fascinating, awe inspiring, a tribute to human creativity and perseverance. Applied in technologies, scientific information creates powerful tools for creative use and devastating misuse. In and of itself, none of this should lead us to think of science as inherently masculine. Yet, because science evolved within patriarchal society, it took on a decidedly masculine tone and became burdened and distorted by a pervasive male bias.

While patriarchal attitudes kept women from prominent positions and full acknowledgement of their abilities and achievements in almost every arena, our society has been particularly discouraging to girls with an interest in, and talent for, science and math (Beckwith & Durkin, 1981; Benbow & Stanley, 1980; Bleier, 1984, Chapter 4; Brophy & Good, 1970; Ernest, 1976; Fennema & Sherman, 1977, 1978; Haven, 1972; Kelly, 1979; Kolata, 1980; Leinhardt, Seewald, & Engel, 1979; Sherman, 1980). Our social system has sought to divide human qualities between men and women, instructing boys that they are naturally intelligent, logical, objective, active, independent, forceful, risk-taking, and courageous. The qualities encouraged in girls have been a different set: sensitivity, emotional responsiveness, obedience, kindness, dependence, timidity, self-doubt, and self-sacrifice. Since an aptitude for science and math clearly implies a bent for analytical intelligence and objectivity, girls have been discouraged from developing their interest in these subjects lest they be considered unfeminine and, thus, socially unacceptable. The personal conflicts so generated steered many women away from math and science and undermined the self-confidence of numerous others who plunged ahead despite societal tracking (Gornick, 1983; Keller, 1977).

In assessing the impact of society's efforts to mold its children, it is essential to realize that *covert, subtle* forces can be exceedingly effective in shaping human behavior. When girls and women are gently discouraged from fully developing their intellectual and creative potential, when they are subtly distracted from seeking positions of power and prestige, the result is the sifting out of all but the most determined minority of women. While the few women remaining in the fray may be cited as evidence that women are not prevented from achieving in our society, actually the probability of a woman succeeding has been drastically diminished by eliminating most female contestants from the field at the start. Those who remain consequently operate in arenas dominated by men, where women are unusual, hypervisible, suspect, frequently patronized, and sometimes ostracized.

Essential to success, moreover, is confidence in one's abilities. As scientists, athletes, artists, or entrepreneurs, we must take risks to have any chance of succeeding. We must wonder and worry whether our intellectual analyses of our projects are sound, whether we have chosen valid and effective technical approaches to achieve our goals, whether our creative inspirations will be greeted with admiration, doubt, or derision. No one, man or woman, can know at the start what will be the outcome of a novel project or a new career, but nothing as effectively brings about defeat as the expectation of defeat. Here society effectively stacks the deck in favor of middle- and upper-class white men. Trained from earliest childhood to imagine themselves as potentially powerful, smart, self-sufficient, inventive, fearless individuals, many men (and very few women) have the expectation of success, the a priori feeling that they can take chances and prevail. These are wonderful tools for coping with panic and despair and the fear of failure. It is a huge benefit that an androcentric society bestows on its male children but this remains largely unacknowledged and unrecognized. Hence, the complacent question continues to be asked, Why *have* there been so few great women scientists, composers, artists, entrepreneurs?

Finally, consider that blatant public success and prestige are unequivocally admired when attained by men but are often problematic for women, who find attaining and holding onto success in conflict with notions about what a woman should be. It is a tribute to the individuality and diversity, the creativity and resourcefulness of human beings, that *any* women succeed under these conditions.

Undeniably then, our society presumes that, because of the personal qualities required, science is an essentially masculine enterprise. The origins and implications of this notion are extensively discussed by Keller (1985). McCormack's wry comment is

> as fashions in the historiography of science change, the qualities considered indispensable to excellence . . . change, but the subordinate, inferior position of women remains the same. . . . In an earlier period when the essential quality of the scientific mind was defined as analytic ability, women were thought to be unintellectual, deficient in reasoning ability. Warm and sensual, they were damned with faint praise for their allegedly "natural" gift of intuitive insight, a desirable but clearly a lower level of skill for the heirs of Descartes. At present when the history of science is being rewritten in terms of creative, Kuhnian (1970) paradigmatic leaps, the brilliant scientific mind is described differently: a type of concentration that is loose, intuitive, a bit frivolous, if not wayward. Women who should be reaping the rewards of this revision are described as being overly cautious, too bound by experimental data, unwilling to speculate and, on the whole, too rational. (McCormack, 1981, p. 2)

While the overwhelming majority of scientists have been men, substantial numbers of women scientists have been productive throughout this century, but they faced layer upon layer of discrimination which, with few exceptions, deprived them of recognition and influence in their fields (American Chemical

Society 1983; Gornick, 1983; Keller, 1977, 1983; National Science Foundation, 1984; Rawls & Fox, 1978; Rossiter, 1982; Sayre, 1975; Vetter, 1980; Watson, 1968; Weisstein, 1979).

In fact, maintaining an army of productive women scientists at the lowest echelons of the profession has been fundamental to the advancement of men scientists, who could take advantage of women's invisibility, immobility, and expectation of self-sacrifice, to claim the research of their subordinates as their own. Thus, women came to be perceived as useful foot-soldiers in science, capable of carrying out the pedestrian laboratory routines that research requires, but lacking the creativity, insights and analytical prowess necessary for innovative research. This is how Rosalind Franklin came to have her extraordinarily fine analysis of the structure of DNA pirated and appropriated by Wilkins, Watson, and Crick, who then turned around and explained to the world that "Rosy" was really good at taking X-ray pictures but would surely not have been capable of interpreting them. Thus, it was all for the best that Wilkins, Watson, and Crick hijacked her data and claimed her discoveries as their own (Sayre, 1975; Watson, 1968).

BEHAVING LIKE A SCIENTIST

With white males holding most scientific posts and all positions with any prestige attached to them, the scientific enterprise itself became fused in people's minds with the character traits (real or imagined) of the typical Western, white, middle-class male. This phenomenon has made it difficult for academic hiring and promotion committees to envision women as suitable colleagues, leading to an uneasiness, which is frequently misattributed to some aspect of the woman scientist's work or personality. An example of a masculine character trait associated with, and expected of, scientists is the drive toward personal power, prestige, authority, and dominion over property and personnel (the more floor space, equipment, technicians, post-docs, and large grants, the better).

Consider the situation young scientists find themselves in as they begin their first tenure-track faculty positions in the science departments of research-oriented universities. As a rule, each new faculty member has about 5½ years during which to demonstrate a sufficient scientific talent and drive to merit a tenured position. A faculty member who fails this test must leave the university. Based only on the theoretical goals and purposes of science, one might suppose that a young scientist had best get to work in the laboratory, developing a research program that shows intelligence, creativity, and originality. One would think that successful young scientists should conduct their research in a careful, well-organized, thorough manner, with energy and persistence, and that they should be reasonably cooperative and completely honest. Naturally, funds and equipment would have to be obtained to develop and sustain such a research program. Graduate students and postdoctoral fellows who became interested in the project would gradually join the research program receiving supervision, support, and assistance.

What often actually occurs is a caricature of the development of a research program, in which form and symbols may become more important than content. The young scientist moves into a laboratory and concentrates on accumulating the largest possible quantity of research funds, instruments, equipment, supplies, and research personnel. Technicians are hired to do the actual research, and potential graduate students are courted and enticed to join this laboratory rather than another. This is done with minimal regard for the personal research interests of the student who, having once given consent, is quickly assigned a subproject within the scientist's research program. When money is available for hiring them, recent PhDs are sought through contacts with friends and former mentors at other institutions, so that postdoctoral fellows will augment the laboratory's research staff and productivity.

Faculty scientists thus take on a sort of chairmanship of their own research-producing corporation, which they then guide and administrate. The products of this corporation (i.e., the research assigned to, and carried out by, the technicians, students, and postdoctoral fellows) are transformed into research articles by the head of the laboratory. This individual is considered the senior author, responsible for the inspiration and intellectual content of the research. To improve one's chances of getting tenure, and one's chances of success in the competition for research funds, as many articles as possible must be produced. This is accomplished by choosing avenues of research that are very straightforward and applying techniques and approaches developed by others, so that lots of publishable data should be quickly forthcoming. These results are then subdivided into small portions, a different article is written about each one, and each article is padded with background information published elsewhere.

Finally, friends and former mentors (the old-boys' network, in other words) are massaged and manipulated to produce invitations for the young scientist to give research seminars at prestigious institutions and conferences. Among friends, this is often a reciprocal arrangement: I'll invite you to speak and then you invite me. It looks good on the record since tenure is granted partly on one's recognition within the scientific community at large.

Once tenure is granted, the pressure decreases only a little. Scientists are anxious to receive further promotions and, if possible, job offers from more prestigious institutions. There is severe competition for research funds, which provide the means to accumulate the symbols of scientific success: new pieces of equipment, more research personnel, trips to numerous national and international meetings. (There is also a tendency to view one's value within the scientific community as equivalent to the sum of the research funds one has been able to attract and this, in turn, influences one's salary.) The idea is to appear big and prosperous, spread your name around. Meanwhile one grows progressively more distant from, and less informed about, the research carried out by others in one's own laboratory (Alberts, 1985).

Scientific research thus becomes an arena of competition for prominence and authority, not unlike the arenas of business and politics. This somewhat grotesque way of organizing a research program has come to be expected of scientists. It is viewed as strong evidence of a scientist's energy, initiative, and ambition; it is taken as a sign of personal drive and future potential. This may create a dilemma for the woman scientist. If she utilizes her male colleagues' methods in the scramble for advancement, she invites criticism, perhaps ostracism, since these intensely competitive behaviors are disconcerting coming from a woman. If, on the other hand, she tends to be docile, helpful and supportive to others, self-sacrificing as a woman is expected to be, she may very well be faulted for not pursuing her career with the appropriate level of vigor and drive.

Fusion of the scientist's image with a masculine authority stereotype is also evident in the public demeanor of scientists. In the patterns of words they choose for use in public lectures and research articles, scientists almost invariably project an image of impersonal authority and absolute confidence in the accuracy, objectivity, and importance of their observations. By all appearances, they will brook neither doubt nor vacillation. This authoritative demeanor is maintained even though it is antithetical to the nature of science, for the data and control experiments that underlie scientific "truth" are always limited (more often than not, just barely sufficient to make the conclusion plausible), the instrumentation and analytical methodology always approximate, and alternative interpretations abundant. The hypothetical, incompletely verified, continually evolving character of scientific "truth" is disguised by a veil of masculine authority. The weaknesses and inaccuracies, the holes in the data, are purposefully hidden as scientists interpose a shield of confident authority between themselves and the public.

It is noteworthy that when women scientists give public lectures about their research, they often call attention to the limitations of their data, to potential flaws in the experimental design, to control experiments that remain to be done. They engage in a kind of public criticism of their own work, taking pains not to overstate their findings or deceive the audience about the work's impregnability. While this approach might be viewed as the woman scientist's effort to be modest and self-effacing in congruence with the stereotype of true womanhood, it is no less a mark of honesty and respect for good scientific practice. Yet, because it diverges markedly from the masculine scientific norm, women scientists who behave this way appear to devalue themselves and the status of their work in the process.

THE STRUCTURE WITHIN WHICH BASIC SCIENCE IS PRACTICED

The average quality of work done in basic research today is, in all likelihood, substantially limited by structural features of modern academic science. Unfortunately, the academic science system evolved in ways that foster and reward rapid publication of multiple research articles, often based on hastily executed research. Of course some excellent, thorough research still is done, but this happens *despite* the selective pressures of the system, and high quality research is overwhelmed quantitatively by superficial, unreliable work that can be churned out much faster. This situation has surely developed in part because judgments on the quality, originality, and thoroughness of research require a substantial investment in time and concentration—an investment too often withheld by university screening committees and peer review panels. So the number of papers published and the superficial characteristics of research are often rewarded instead of quality.* There is, for example, a tendency to uncritically adopt as valid and standard-setting, the theoretical and experimental approaches emanating from a small number of highly prestigious laboratories, as well as their research results. Then, other scientists, who adopt the trendy methodologies and report the expected research results, receive accolades and ready access to the professional journals, while scientists using original approaches, exploring novel territory or obtaining unexpected (sometimes challenging and unwelcome) results, encounter obstacles to acceptance and publication. Such emphasis on fashion and getting the expected results, accompanied by a suspension of critical judgment, encourages the accumulation of poorly controlled, unreliable research, and the consequent entrenchment of incorrect conclusions.

Another structural problem in academic science is the excessive and destructive level of competition. While competition often is effective in augmenting motivation and dedication to one's scientific career, it is antithetical to a fundamental characteristic of science—the need to share one's methods and results. Each experiment is based on the previous discoveries of others. Each result must be validated by testing in altered circumstances in other experimental settings. Yet, because of the severity of competition, many scientists are fearful of discussing their ideas and methods with others, who might quickly exploit this information for their own gain. In fact, in a disquieting number of documented instances, prominent scientists, acting as reviewers, advised journal editors not to publish a submitted article or advised a grant-

---

**Editor's note*: It is important to note, however, that the pressures described here vary considerably among scientific fields, among science departments in universities, and among scientists. There are scientists and departments that recognize and value quality and originality rather than quantity and faddishness.

ing agency not to fund a body of proposed research, then quickly carried out the same project in their own laboratories and published it under their own names (Broad, 1980). The fear of being scooped further promotes hasty publication of insufficiently developed research, so as to establish priority of authorship. The excessive pressure to publish also promotes outright fraud, the fabrication of data never collected, which has become a substantial problem in science (Broad, 1980, 1982; Broad & Wade, 1983; Culliton, 1983; Hunt, 1982; Rensberger, 1977; Smith, 1985c).

Under these various influences, the professional literature has swelled to gargantuan proportions, making it extremely time-consuming just to keep up with developments in one's own specialty. Analyzing pertinent articles with care has become next to impossible because there is far too much to read. Excessive, mind-numbing specialization is another consequence, as scientists attempt to cope with the out-sized literature by reading only articles most directly affecting their own experiments.

Most scientists understand that much of what is turned out in research is prematurely published and not of the highest quality. The pressures of the system leave scientists little choice. Heroes and martyrs being few in number, the academic science system too often generates mediocrity, conformity, tension, fear, alienation, depression, and wasted energy.

THINKING LIKE A SCIENTIST

The predominance of white, middle-class men in science research has resulted in excessive reliance on conceptual paradigms related to the social preoccupations of this group. Consequently, research priorities have tended to become distorted, as have the design and interpretation of experiments. (For a lucid review of this subject, see Bleier, 1984.)

The effect of male bias on scientific research is most dramatic in fields closely related to sex and gender. In the study of primate social behavior, for example, scientists have exaggerated the extent and importance of male dominance hierarchies and male aggression, initiative, and competition in controlling troop behavior (Haraway, Hrdy, this volume; Leibowitz, 1979; Wasser, 1983). This astigmatism has seriously compromised data collection and theory construction in animal behavior and evolution. Another example of distortion can be found in the near-obsessive focus on discovering a biological basis for small average differences between the sexes in behavior, or in the scores achieved on some kind of cognitive test (Bleier, 1984, and this volume). Thus, in areas of biological research related (sometimes subconsciously) to human sexuality or gender, male bias in the scientific establishment has frequently resulted in misleading, unreliable research. One might suppose that the value scientists place on "objectivity" would lead them to guard against such distortions, but in traditional scientific practice one is trained to devise "control" experiments to detect the possible influence on observations of all factors *except* cultural bias.

Masculine cultural bias in the design and interpretation of experiments also influences fields further removed from areas of sex and gender, but here the distorting effects are likely to be less blatant. After all, numerous factors compromise the effort to maintain a level of scientific objectivity commensurate with accuracy and realism in research. Among these influences are personal ambition, faddishness in research design and data interpretation, the pressure toward rapid publication, and a plethora of culturally derived presuppositions. With many influences acting simultaneously, when research becomes distorted, numerous factors might be suspected of contributing to the problem. This should provide historians and philosophers of science with a productive new focus for research. (For example, see the thoughtful analysis in Keller, 1985, chapters 7, 8, and 9.) Let me suggest two research areas in which a preoccupation with dominance and control has tended to distort biological research at the cellular and molecular level.

An example of a gender-related paradigm that has long influenced biology is the concept of the living organism or cell as a machine controlled by a master molecule (DNA, the material of which genes are composed) containing coded information that is progressively revealed by a centralized readout program. In its extreme form, this model envisions the cell or organism as the passive, subordinate recipient of directions and orders from the master, the repository of the coded information without which nothing would get done. As the king to his subjects, as the architect to his builders, as upper management to lower labor, so stands the collectivity of the organism's genes (the genome) to the cell in which it resides. As Goldschmidt stated in 1940

> the specificity of the germ plasm [equivalent to genome in his usage] is its ability to run the system of reactions which make up the individual development, according to a regular schedule which repeats itself . . . with the purposiveness and orderliness of an automaton.

When transferred into the field of animal behavior, these themes of masculine authority, intellectual/rational domination and control sound like this:

> The females were incapable of "governing" the group and social tension and disorganization were constant. The introduction of but one adult male into the group corrected the situation immediately, and a more normal political and social pattern quickly returned . . . primate females seem biologically unprogrammed to dominate political systems . . . (Tiger, 1977, p. 28)

The simplified notion of the genome as director of the cell's and organism's activities has been useful in analyzing gene expression, particularly in bacteria. Distorting simplifications can be useful temporary devices because they break down complex reality into portions small enough to grasp intellectually and manipulate in experiments. Yet, the simplistic notion of the genome as sole information source and dictator of the organism's activities resonates a little too well with white, middle-class male concepts of power, authority, and dominance. What began as a conceptual simplification was soon mistaken for reality and imposed on complex systems, where the fit was poor. Conflicting evidence tended to slip from people's attention, and research priorities became distorted.

The analysis of early embryonic development is an area where theories of the primacy of the genome fit poorly. Although the genome specifies the primary structure of the enzymes and structural proteins essential to development, the patterns of oogenesis (egg development) and early embryonic differentiation clearly depend on *when* and *where* each of these proteins is *utilized* in the egg or early embryo, factors that are largely beyond the reach of the genome (Jeffery & Raff, 1983; Malacinski & Bryant, 1984; Raff & Kaufman, 1983). Thus, in contrast to the situation found in bacteria, a long interval often elapses between read-out of genetic information in eggs and embryos and the utilization of the resulting protein. Moreover, the products of different genes are stored in the egg cytoplasm for different lengths of time before becoming active. What decides the time of activation of gene products is largely a mystery, but it is likely to be one of the keys to understanding how embryos develop into organisms. Another essential factor in the organization and development of embryos is the spatial arrangement of molecules. In different embryonic cells, each of which contains an identical and complete copy of all the organism's genes, a different subset of the genes is expressed. Nobody knows what brings this about, but it certainly isn't the genes themselves. As a result, different proteins are synthesized in different regions of the embryo, a feature essential to development and to the differentiation of different groups of cells into different tissues and organs. Once synthesized, each protein becomes localized to a specific subcellular place within the egg or embryonic cell, another essential feature of development that depends, at least in part, on factors beyond the genome.

Yet in the early 1960s, when most biologists became fully aware of the key role played by genetic transcription (read-out of genetic information) in controlling the characteristics of bacteria, there was a massive shift among developmental biologists toward research in genetic transcription during embryonic development (and related protein synthesis) to the exclusion of other avenues of embryological research, which suddenly seemed hopelessly old-fashioned and wrong-minded. As we contemplate the fruits of this massive research concentration on gene expression during embryonic development from our vantage point in 1985, it seems as though we learned quite a lot about gene products and their time of synthesis during development, but we learned remarkably little about how embryos become organized to make development of new structures possible. We became so mesmerized by the concept of the "Master Molecule" that we overlooked the obvious fact that, since the time that the first cells evolved several billion years ago, genes have acted exclusively within a highly structured, information-packed, elaborately differentiated cell. Neither in nature nor in test tube experiments does the genome show any ability to direct the formation of a cell or an organism. Each cell arises from a preexisting *cell*; that is, from an entity orders of magnitude more complex than the organism's genome. As we drifted through the 1960s and 1970s, propelled on currents of research fashion, we mistook our simplified working models for real eggs. Our judgments were a bit cloudy when we chose what to work on, so we didn't accomplish as much as we might have.

An analogous situation existed during the 1950s and 1960s with respect to the mechanism underlying the phenomenon known as *primary embryonic induction*. In the 1920s, Hans Spemann, together with his collaborator Hilde Mangold, reported that in amphibian embryos the brain and spinal cord develop as a consequence of stimulation of the ectoderm by the underlying notochord (Gilbert, 1985, pp. 254–268; Spemann & Mangold, 1924; Spemann, 1938). In a series of elegant experiments, Spemann and Mangold demonstrated that the nervous system would develop from whatever part of the ectoderm came in contact with notochordal tissue. The notochord was consequently designated as the *inducer* while the ectoderm was thought of as the responding tissue. If induced, the ectoderm developed into brain and spinal cord; if not induced, the ectoderm would develop into epidermis.

In subsequent years, many inductions were found to occur during embryonic development. Each follows the same pattern: Tissue A develops into Structure A if Tissue B first induces it by coming into close proximity for a time. But it also became apparent that a great variety of tissues, and tissue extracts from a variety of organisms, could induce amphibian ectoderm to develop into brain and/or spinal cord without the participation of the notochord. In fact, ectoderm alone, when isolated in vitro and subjected to various kinds of sublethal toxic physical and chemical stimuli, proved capable of developing into brain and spinal cord. All of this should have made it clear that ectoderm was fully equipped to develop into brain and spinal cord on its own, and that the inducer probably transmits no more than a generalized signal, which might influence the precise timing and spatial location of this developmental event.

Yet, the paradigm of a directorial inducing molecule imparting an essential piece of information to a dependent, receiving tissue was too intoxicating to abandon. Consequently, fruitless research projects were pursued for years, the object of which was to trap the inducing molecule in a filter placed between the inducing and responding tissue, then to perform a laborious biochemical analysis of the material in the filter, testing each molecular component for its ability to induce ectoderm to form brain and spinal cord. As might have been anticipated, no consistent results emanated from any of these experiments. Thus, fashion and foolishness, exacerbated by too strong an attachment to a male-authority paradigm, appears to have derailed productive analysis of this topic for a time.

THE SCIENTIST AS INGENUE

Despite the fevered atmosphere of many modern research laboratories, scientists usually envision themselves pursuing their research in an abstract sphere, isolated from political and cultural influences, much as the Austrian monk, Gregor Mendel, tabulated the inherited characteristics of the peas he grew in his 19th century monastery garden. Isolation from political and cultural influence is thought to be essential for achieving scientific objectivity. Scientists think of themselves as totally rational, neutral beings who have no political agenda and neither interest in, nor responsibility for, the ways in which their research is interpreted or utilized by society.

The scientific mind and the scientific method are thought to ensure the neutrality and objectivity of scientific research, and of scientists' pronouncements. All scientists need to do is to steer clear of political and social movements that could undermine their objectivity. With this mind-set, scientists believe that any political movement, like feminism, that seeks to influence science would undermine the essential neutrality of the scientific enterprise.

Yet science has not been neutral. Repeatedly, in the course of history, the pronouncements of scientists have been used to rationalize, justify, and naturalize dominant ideologies and the status quo. Slavery, colonialism, laissez-faire capitalism, communism, patriarchy, sexism, and racism have all been supported, at one time or another, by the work of scientists, a pattern that continues unabated into the present (Bleier, 1984, and this volume; Goldberg, 1974; Gould, 1981; Hubbard & Lowe, 1979; Lewontin & Levins, 1976; Lewontin, Rose, & Kamin, 1984; Merchant, 1980). In truth, scientists are no more protected from political and cultural influence than other citizens. By draping their scientific activities in claims of neutrality, detachment, and objectivity, scientists augment the perceived importance of their views, absolve themselves of social responsibility for the applications of their work, and leave their (unconscious) minds wide open to political and cultural assumptions. Such hidden influences and biases are particularly insidious in science because the cultural heritage of the practitioners is so uniform as to make these influences very difficult to detect and unlikely to be brought to light or counterbalanced by the work of other scientists with different attitudes. Instead, the biases themselves become part of a stifling science–culture, while scientists firmly believe that as long as they are not *conscious* of any bias or political agenda, they are neutral and objective, when in fact they are only unconscious.

CURRENTS FOR CHANGE

Changes in the expectations and status of women in the United States over the past 15 years, brought about by the feminist movement, have made entry- and middle-level positions in academic science available to women for the first time in history. Small numbers of women are being admitted to the faculties of major research-oriented universities. As they begin teaching in their fields, developing their own research programs, and contributing their ideas at national and international meetings of their professional organizations, these women have the potential to broaden perspectives within the men-only preserve that has been science. Their presence could ultimately have an impact on the goals and values of scientists and on the structure of academic science. Maybe.

The admission of women in symbolic numbers to positions in academic science is a far cry from according them the opportunities and attention requisite for real impact. *Overt* sex discrimination has become a bit unseemly in academe, but its fraternal twin, *covert* sex discrimination, thrives within academic science. What was remarkable about the recent scurrilous remarks about women scientists by "Our Father of DNA" was not that a prominent scientist harbored these beliefs, but only that he gave voice to them in public (Bleier, this volume; Culliton, 1985). Thus, in science, as in society at large, a devastating legacy of sex discrimination remains, a pervasive tradition of discounting the opinions, values, and abilities of women. This will not be overcome quickly.

Moreover, the tenuous toehold, which is all that women have within science, serves to straitjacket, even to silence, women's voices. As scientists who are women, we start with a substantial handicap, a "prove it to me that you're OK" stance on the part of our peers. To gain acceptance we must demonstrate that there is nothing unusual, no deviation from the norm (men), in our attitudes and beliefs. This gives us an enormous incentive to impersonate men scientists. It tends to make women scientists conservative in their work and in their publicly expressed, even privately held, beliefs.* As Keller (1982, p. 590) noted, women, like other "outsiders," tend to "internalize the concerns and values of a world to which they aspire to belong."

**Editor's note*: There have always been outspoken exceptions to this generalization, and today increasingly articulate feminist voices are being raised within the scientific disciplines. For many years, there have been women's caucuses within chemistry and the neurosciences, and AWIS (American Women in Science) has existed for about 10 years. The July 1985 issue of the *Women in Neuroscience Newsletter* carried two articles written by members (Leanna Standish and Eve Andersen) urging members to read the recent feminist critiques of science and calling for a feminist retheorizing and restructuring of the neurosciences. Similarly, the June 1985 issue of the *Society for the Study of Reproduction Newsletter* has an article by Rita Basuray introducing

For the time being, therefore, pressures arising *outside* of science, from within the political feminist movement, promise to be more important in influencing the direction of scientific thinking, than the stated opinions of women scientists themselves. Gradually, though, as feminist currents lap against the foundations of science, impinging on the barnacles, progressive and feminist attitudes will come to seem less peculiar, less arcane, perhaps less threatening to scientists. Progressive/feminist women and men will become audible and visible among other scientists and will have a chance to influence how scientists think. (For a discussion of how science is being altered by feminism see Rosser, 1985, and Chapter 8.)

There is, at present, one area within biology where feminist pressures working alongside the creative energies of women scientists have already wrought a substantial reappraisal. This is the field of primate behavior, as noted in the introduction to this volume and described by Hrdy in Chapter 6. Assumptions about gender differences in *human* society had served to blind investigators to the active, diversified, individually adaptive, aggressive, dominating behaviors of female primates. Considered unacceptable and unnatural for females in *human* society, these behaviors were ignored, denied, and made to vanish from the data on *primates*. But as the feminist movement expanded our notions of women's roles and abilities, some ethologists (primarily women) became aware that the field data they were collecting about primate societies did not support the traditional notions of male dominance and control abetted by female passivity and self-sacrifice thought typical of all primate societies (Hrdy, Chapter 6). It gradually became clear that deeply held beliefs, coupled with poorly designed techniques of data collection and analysis, had led the science establishment into grave distortions in the analysis of primate sex roles and social organization, which were then used, in turn, to support sociobiological notions about the inevitability of patriarchal social organization in *human* society (Bleier, 1984, Chapters 2 and 5; Goldberg, 1974; Hrdy, 1981; Wasser, 1983).

That our earlier concepts of primate social organization were gravely distorted is now widely acknowledged. The evolution of our ideas about primate behavior illustrates vividly how scientific investigations are influenced by social forces and by scientists' preconceptions. After all, "reality" is much too diverse and complex to be analyzed by scientists. To make scientific study feasible, scientists focus on a tiny subsection of reality, chosen to be small

---

the idea to members that "science is a social activity subject to a variety of cultural pressures" and discussing the question of male bias. Thus, caucuses established to protect and further women's equal rights to careers within traditionally male fields are paying increasing attention to the problems that are being addressed by feminist writers in the theoretical/methodological frameworks of scientific disciplines. This is the kind of evolution identified by Sue Rosser in Chapter 8.

enough and simple enough to submit to rational analysis. What subfraction of reality we choose to study, what questions we decide to ask of it, what methods we apply to our analytical task, and what ideas we bring with us when we interpret and evaluate data are all profoundly influenced by our personal and societal views, our values, our language, our concerns.

"Scientific objectivity," by which I mean the effort to apprehend and understand an empirical reality uncontaminated by our personal preconceptions, is exceedingly difficult to approach and impossible to attain. Yet, the more unsubstantiated assumptions we can eliminate or verify empirically, and the more underlying presuppositions we become cognizant of, the more profound our scientific understanding becomes. But the greater the uniformity in the gender, culture, and class background of scientists, the more remote become the possibilities for profound insight and an approach toward greater objectivity. On the other hand, when there is gender, culture, and class diversity among scientists, we can challenge each other's ideas and assumptions. Together, we can identify many of the hidden presuppositions that distort a scientific study.

A fascinating debate continues among feminists interested in science, concerning just how science would be different if women played a large role in it (Keller, 1982, 1985; Bleier, Introduction to this volume). Some feminists believe that a science designed and practiced by women would be radically different from today's science because women's values and goals are so different from men's. They note that the traditional scientist tends to assume a cold, detached, exaggeratedly intellectual attitude vis-à-vis his object of study, a stance that is alien to women's behavior. Moreover, Western science is said to be inextricably bound up with the purposes of dominating and manipulating nature, purposes held to be of less importance to women. In a science designed by women, so the argument goes, subjects would tend to be viewed in their larger context, with greater attention to the linkages between different levels of organization, and between different aspects of the same subject. It is the WHOLE, complete with all its details and idiosyncracies and individualities that is important in a female worldview. Abstraction, reductionism, the determination to repress one's feelings to promote "objectivity," have not the same priority to women as they do to men (Griffin, 1978; Keller, 1985; Marks & de Courtivron 1980; Merchant 1980; Rose, this volume). Other feminists doubt that women's worldviews are so radically different from men's. They see little in scientific thinking or in the objectives of science that is alien to working women scientists. So these feminists expect moderate, rather than radical, differences between a science influenced by feminism and patriarchal science (Gornick, 1983; Keller, 1982, 1983).

I share the latter view because I am struck by the many different ways that we can use our minds to gain understanding. The procedures of science are only one of these. By striving for objectivity, repeatability, and verifiability of every fact and every generalization, scientists ferret out, then catalogue, the properties that all members of a class of things have in common. What is idiosyncratic comes close to defying scientific analysis because the scientific method of study presupposes that we can examine many instances of the same thing. Only then can we identify regularities. Only when we have many repeating instances can we do experiments. Since each creature, each event, is unique when considered in its totality, in all its details, in its context, science must abstract certain repeatable properties of things in order to have material suitable for scientific procedures. The knowledge that results is remarkably reliable and useful, not to mention interesting, but it is limited to the characteristics things share *in common*; the individual is excluded. The taste and smell and feel of full reality is jettisoned to preserve the general and predictable.

Poetry and painting, in contrast, dwell in the particular. They can convey things that are generally true, but only through the particular instance. The images created by poets and painters may be viewed as more complete and profound, truer to the essentials of life, than scientific generalizations, but the latter are more reliable, easier to draw conclusions from, and therefore especially useful for getting things done. Human behavior, according to this view, might be better understood through literature and poetry than through scientific analysis, because human behavior is so very particular, so very variable, so reactive to context as to threaten the possibility of generalization and prediction.

Science is inordinately valuable and deeply interesting; yet, it produces only one type of knowledge. I expect that science, literature, poetry, music, architecture, philosophy, history, and so forth will all be influenced, perhaps even transformed, by feminism, but each will retain its distinctive way of using the mind and the spirit. In the sections that follow I sketch the changes I would like to see feminism bring to science, and the type of scientific practice that might result.

IN WHICH SCIENTISTS REJOIN SOCIETY

Science has always been embedded in cultural history, influenced by political, economic, and social forces operating in society at large. Powerful groups have repeatedly harnessed scientific research to serve their own interests, and individual scientists have appeared in public as their apologists (Bleier, 1984; Gould, 1981; Lewontin et al., 1984; Rose & Rose, 1976). Let me cite just a few of many possible examples.

Corporations that thought they could reap huge profits by constructing nuclear power plants sought to garner public support by citing evidence provided by nuclear physicists, chemists, and geneticists on the efficiency, feasibility, and safety of the nuclear systems for power generation. In the course of the numerous public and legislative hearings which ensued, scientists have been called upon not only to provide facts and explanations, but also to evaluate feasibility, safety, and the cost–benefit ratios of alternative reactor designs and safety systems. Moreover, when contemporary scientists design research projects, they do so with one eye focused on the objectives and attitudes of the government agencies and private organizations that might be asked to fund the research. Since it is the already powerful in society who fund research, and who can influence the direction of government funding, scientists' efforts are funneled into research activities that primarily benefit the overprivileged.

Another example. Early in this century, groups within the United States who attempted, with success, to get Congress to set limits on immigration from Eastern and Southern Europe, received strong support from geneticists who testified that a large proportion of the immigrants from Poland, Yugoslavia, Greece, Italy, and so forth, were demonstrably "feeble-minded," as evidenced by intelligence tests administered to them upon their arrival in the United States (Gould, 1981, Chapter 5; Kamin 1974).

Finally, it may be noted, at the present time in the United States and Western Europe, conservative political groups who oppose social welfare programs and equal opportunity legislation draw on the research and stated opinions of geneticists and sociobiologists who proclaim that human behavioral differences are largely determined by genes, so that remedial improvements in education, economic circumstances, etc., would be expected to have only a minor effect (Lewontin et al., 1984).

Now, little of this should come as a surprise. It is quite obvious that interest groups will seek to strengthen their arguments with evidence provided by cooperative scientists, particularly considering the esteem science enjoys in modern Western societies. Moreover, it is perfectly natural that the worldview, suppositions, conceptual frameworks, and patterns of thinking of scientists would be substantially shaped by the culture within which scientists mature and receive their education.

What is truly remarkable is that scientists and the public deny this. Scientists and the information they collect are treated as though they are culture-free, classless, apolitical; as though the scientist's attempts at objectivity were routinely successful. While the statements of a lawyer, a union organizer, a corporate executive are sifted and weighted relative to the speaker's probable motivations and prejudices and the listener's own views, the pronouncements of scientists are treated with reverence. They may be ignored but they are rarely challenged. (Even Papal pronouncements receive more spirited public debate!)

My contention is not that scientific evidence is of little value or deserves to be ignored when decisions are made. Scientific evidence is extremely valuable when considered relative to the research design, methodology, and meticulousness that went into collecting the information. Yet the reliability of scientific judgments and conclusions is seriously compromised because scientists commonly deny (to themselves no less than to others) that their research designs, methodologies, and interpretations may be infiltrated by cultural and social biases. So nothing is done to detect or prevent this. Knowledge of the history and sociology of science, among scientists as well as the public, could do much to stem the misunderstanding and misuse of scientific authority.

I would like to see scientists confront head-on the probability of cultural bias and distortion in choice of research problems, use of language for formulating and describing what ought to be studied, choice of experimental design, methods of data collection and analysis, and the evaluation and interpretation of the results. All these should become matters for introspection, for critique and discussion within research groups, and for open acknowledgment and analysis within the discussion sections of research papers. This process would be facilitated by involving culturally diverse people in the analysis and evaluation of research.

Above all, these scientific quality control circles should consider whether we are not being led astray by scientific paradigms and modes of analysis that fit poorly with the evidence at hand. For example, are we assuming the importance of dominance hierarchies where they might not exist at all? Are we assuming a genetic or other biological cause where cultural influences are likely to be decisive (e.g., in determining intelligence, schizophrenia, criminality, entrepreneurial energy, homosexuality, aggression)? Is our reductionist approach to our problem introducing undue distortion, so that we are being led away from understanding rather than towards it? Have we given sufficient consideration to a variety of theories that might explain our observations, or have we excluded workable hypotheses without adequate justification (Chamberlin, 1965)?

IN WHICH SCIENTISTS TAKE REASONABLE RESPONSIBILITY FOR THEIR WORK

Science is a powerful tool for good as well as for evil, for emancipation as well as for exploitation. How scientists use their time and talent, their training at public expense, their public research funds, and the public trust are not matters to be brushed aside lightly. Many fascinating research problems wait for attention. Residing, as we do, inside a universe filled with enigmas, many of which lend themselves to research approaches congenial to our personal styles, many with applications beneficial to segments of society that are due for some benefits, how do we justify working on research whose applications threaten to be deeply destructive of natural resources, of human life, of the dignity and self-respect of a racial or ethnic or gender group?

As I sit here composing these words I can hear the more tradition-bound scientists moaning: "Heaven help us! That would be the end of legitimate science. Harness science to social or political purposes and it will be totally destroyed."

Well, obviously science is being heavily exploited for political and economic and medical gain at this very moment. As this page rolls through the typewriter, a segment of the scientific community is lining up at the Strategic Defense Initiative funding window to collect financial support for basic research that will make it possible to carry on wars in space (Norman, 1985a, 1985b; Smith, 1985a, 1985b). Another group of scientists is waiting at the mechanical heart window to receive funding for the design and improvement of devices that will serve a tiny minority of highly privileged men. Undoubtedly, interesting and useful knowledge will flow from these projects but is this the best way to do science?

In sum, as scientists and as human beings, we are obliged to make responsible choices about what we do in our work. We must be knowledgeable about how our research is likely to be applied, and do what we can to prevent dangerous, detrimental applications while promoting beneficial ones. Furthermore we must cope responsibly with the by-products of our research designs: for example, radioactive and chemical pollution; the danger that harmful microbes could be inadvertently released; the possibility that our experiments cause needless suffering among experimental animals, because the results are not important enough to warrant any suffering at all or because we could have designed the study utilizing other organisms or procedures that would reduce or avert the suffering.

Finally, scientists must accept some responsibility for the ways in which their research is communicated to the public. By taking the time to educate reporters and science writers about the background, implications, and limitations of one's research, the scientist can do much to promote accurate, realistic descriptions of research in the media. Scientists who permit their work to be presented publicly in a distorted, overblown, incomplete, misleading manner are poor scientists, even allowing for the difficulties of dealing with the media. Science reports need not forever mislead, frighten, and alienate the nonscientist, or disadvantage the underprivileged, or simply serve to confirm in the minds of many "that their social prejudices are scientific facts after all" (Gould, 1981, p. 28).

IN WHICH WE REDEDICATE OURSELVES TO QUALITY AND HONESTY IN RESEARCH

We must redirect the science establishment to reward thorough, thoughtful, honest, and cooperative researchers while penalizing scientists who are dishonest, self-serving, and careless of what they do in the name of science.

The structure of the research and publications system within which scientists work and advance must be modified to favor high-quality, thoroughly researched pieces of work published, if necessary, at longer intervals. While bloated publication lists are important to advancement in contemporary science, they really should be grounds for suspicion that the research is being done in a shoddy or superficial manner. Bandwagonism in research focus and excessive preoccupation with trendy instrumentation should be viewed with appropriate levels of skepticism.

We must humanize the science curriculum and the scientists. Courses in ethics and in the history and sociology of science must become part of the normal science curriculum, particularly important for graduate students and medical students, who will become working scientists. Graduate programs must expect and reward honesty and openness about the limitations and uncertainties of one's experiments, while penalizing bluff and bluster. We must strive for more humane working conditions for all of science's practitioners; conditions that encourage thoroughness, honesty, and care in one's work.

EPILOGUE

Science has great influence within Western societies for three reasons. First, the knowledge it provides translates easily into technological and medical advances that are widely desired. Second, science exemplifies some important societal values such as intelligence, rationality, dispassionate objectivity, perseverance, dominance, and control of nature. Third, it satisfies the human craving for a worldview, for a means of understanding the universe and our position in it.

Although scientists generally regard "reality" as fixed, immutable, always out there, unaltered by the scientific enterprise itself, it is more truthful to view reality as partially created by science (Knorr-Cetina, 1981; Latour & Woolgar, 1979). Since only minute fragments of reality are examined by scientists in any particular generation, the choice of scientific focus, and the modes of thought chosen for scientific analysis, do much to create a partial reality for the rest of society. Then, when scientific information is transformed into technological or medical tools, this alters reality even further from what it would have been with a different type of science. Viewed in this light, science has profound effects on society and cultural evolution. If feminism succeeds in modifying science, the changes will reverberate in the larger society.

Some readers may wonder whether, in reconceptualizing science from a feminist perspective, we are not going off the track when we find ourselves promoting views that a majority of progressive men would probably support. In my view, definitely not. Women and men share virtually all abilities, characteristics, and attitudes. The differences between men and women are matters of emphasis and concentration, not a question of absence versus presence. Consider characteristics such as competitiveness, aggression, the desire to dominate, hunger for recognition, a predilection for analytical thinking, or for intuitive/creative synthesis, cooperative workstyle, alacrity and resourcefulness in the field or at the workbench, pride and creativity in instrumentation, dogged perseverance. Clearly the variability *among* men and *among* women in these diverse traits is much greater, *despite* cultural pressures regarding proper gender roles, than is the difference *between* women and men. Moreover there is every reason to think that the psychological average differences that we note between women and men today result largely, perhaps entirely, from acculturation and so are subject to cultural evolution. As Bleier (1984, p. viii) notes:

> it is not our brains or our biology but rather the cultures that our brains have produced that constrain the nearly limitless potentialities for behavioral flexibility provided us by our brains.

Feminism promises to revitalize and improve the practice of science for women and men alike, as feminist reformers have revitalized and enriched our institutions again and again in history, each time profoundly affecting conditions for both women and men. What Sojourner Truth and other women crucial to the Abolitionist Movement did for ethical standards in this country, what Rachel Carson did for ecological awareness, what Mary Elizabeth Garrett and her associates did for medical education standards, feminist scientists will do for science. Patriarchal science needs a coronary bypass, and feminism is providing it.

Acknowledgements—The sounder ideas in this chapter were developed in discussions within the *Women in Science Study Group* at the University of Wisconsin, Madison. The members included Ruth Bleier, Sheila Burke, Cindy Cowden, Virginia Cox, Diane Derouen, Freja Kamel, Tricia Kelleher, Monica McCarthy, Susan Millar, Charlotte Otten, Linda Sabatini, Marsha Segerberg, Mariamne Whatley, Patricia Witt, and Sandy Zetlan. To each of them, as well as to Judy Enenstein and Val Woodward of the University of Minnesota, and Sue Rosser of Mary Baldwin College, I am deeply grateful for their incisive and thought-provoking comments.

REFERENCES

Alberts, B. M. (1985). Limits to growth: In biology small science is good science. *Cell, 41*, 337–338.

American Chemical Society. (1983). Medalists study charts women chemists' role. *Chemistry and Engineering*, Nov. 14, 53.

Beckwith, J., & Durkin, J. (1981). Girls, boys and math. *Science for the People, 13*, 6–35.

Benbow, C., & Stanley, J. (1980). Sex differences in mathematical ability: Fact or artifact? *Science, 210*, 1262–1264.

Bleier, R. (1984). *Science and gender: A critique of biology and its theories on women.* Elmsford, NY: Pergamon Press.

Broad, W. J. (1980). Imbroglio at Yale (I): Emergence of a fraud. *Science, 210*, 38–41.

Broad, W. J. (1982). Report absolves Harvard in case of fakery. *Science, 215*, 874–876.
Broad, W. J., & Wade, N. (1983). *Betrayers of the truth*. New York: Simon and Schuster.
Brophy, J. E., & Good, T. L. (1970). Teachers' communication of differential expectations for children's classroom performance: Some behavioral data. *Journal of Educational Psychology, 6*, 365–374.
Chamberlin, T. C. (1965). The method of multiple working hypotheses. *Science, 148*, 754–759.
Culliton, B. J. (1983). Coping with fraud: The Darsee case. *Science, 220*, 31–35.
Culliton, B. J. (1985). Watson fights back. *Science, 228*, 160.
Ernest, J. (1976). *Mathematics and sex*. Santa Barbara: Mathematics Department, University of California.
Fennema, E., & Sherman, J. (1977). Sex-related differences in mathematics achievement, spatial visualization, and affective factors. *American Educational Research Journal, 14*, 51–71.
Fennema, E., & Sherman, J. (1978). Sex-related differences in mathematics achievement and related factors: A further study. *Journal for Research in Mathematics Education, 9*, 188–203.
Gilbert, S. F. (1985). *Developmental biology*. Sunderland, MA: Sinauer Associates.
Goldberg, S. (1974). *The inevitability of patriarchy*. New York: Morrow.
Goldschmidt, R. (1940). *The material basis of evolution*. New Haven, CT: Yale University Press.
Gornick, V. (1983). *Women in science: Portraits from a world in transition*. New York: Simon and Schuster.
Gould, S. J. (1981). *The mismeasure of man*. New York: W. W. Norton.
Griffin, S. (1978). *Woman and nature: The roaring inside her*. New York: Harper and Row.
Haven, E. W. (1972). Factors associated with the selection of advanced academic mathematical courses by girls in high school. *Research Bulletin*, 72–12.
Hrdy, S. B. (1981). *The woman that never evolved*. Cambridge, MA: Harvard University Press.
Hubbard, R., & Lowe, M. (1979). *Genes and gender II*. New York: Gordian Press.
Hunt, M. (1982). A fraud that shook the world of medical research. *Medical Economics*, Feb. 15, 207–224.
Jeffery, W. R., & Raff, R. A. (Eds.). (1983). *Time, space and pattern in embryonic development*. New York: Alan Liss.
Kamin, L. (1974). *The science and politics of I.Q.* Potomac, MD: Erlbaum.
Keller, E. F. (1977). The anomaly of a woman in physics. In S. Ruddick & P. Daniels (Eds.), *Working it out*. New York: Pantheon Books.
Keller, E. F. (1982). Feminism and Science. *Signs, 7*, 589–602.
Keller, E. F. (1983). *A feeling for the organism: The life and work of Barbara McClintock*. New York: W. H. Freeman.
Keller, E. F. (1985). *Reflections on gender and science*. New Haven, CT: Yale University Press.
Kelly, A. (Ed.). (1979). *The missing half: Girls and science education*. Manchester, England: Manchester University Press.
Knorr-Cetina, K. D. (1981). *The manufacture of knowledge*. Elmsford, NY: Pergamon Press.
Kolata, G. (1980). Math and sex: Are girls born with less ability? *Science, 210*, 1234–1235.
Kuhn, T. S. (1970). *The structure of scientific revolutions*. Chicago: The University of Chicago Press.

Latour, B., & Woolgar, S. (1979). *Laboratory life: The social construction of scientific facts*. Beverly Hills, CA: Sage.
Leibowitz, L. (1979). "Universals" and male dominance among primates: A critical examination. In R. Hubbard & M. Lowe (Eds.), *Genes and gender II*. New York: Gordian Press.
Leinhardt, G., Seewald, A. M., & Engel, M. (1979). Learning what's taught: Sex differences in instruction. *Journal of Educational Psychology, 71*, 432–439.
Lewontin, R. C., & Levins, R. (1976). The problem of Lysenkoism. In H. Rose & S. Rose (Eds.), *Ideology of/in the natural sciences*. Cambridge, MA: Schenkman Publishing Company.
Lewontin, R. C., Rose, S., & Kamin, L. J. (1984). *Not in our genes: Biology, ideology and human nature*. New York: Pantheon Books.
Malacinski, G. M., & Bryant, S. V. (Eds.). (1984). *Pattern Formation*. New York: Macmillan.
Marks, E., & de Courtivron, I. (Eds.). (1980). *New French feminisms: An anthology*. Amherst: University of Massachusetts Press.
McCormack, T. (1981). Good theory or just theory? Toward a feminist philosophy of social science. *Women's Studies International Quarterly, 4*, 1–12.
Merchant, C. (1980). *The death of nature: Women, ecology and the scientific revolution*. New York: Harper and Row.
National Science Foundation. (1984). *Women and minorities in science and engineering*. Report 84-300.
Norman, C. (1985a). Pentagon seeks to build bridges to academe. *Science, 228*, 303–305.
Norman, C. (1985b). DOD program proves attractive. *Science, 228*, 1291.
Raff, R. A., & Kaufman, T. C. (1983). *Embryos, genes and evolution*. New York: Macmillan.
Rawls, M., & Fox, S. (1978). Women in academic chemistry find rise to full status difficult. *Chemical and Engineering News*, Sept. 11.
Rensberger, B. (1977). Fraud in research is a rising problem in science. *The New York Times*, Jan. 23.
Rich, A. (1977). Introduction. In S. Ruddick and P. Daniels (Eds.), *Working it out*. New York: Pantheon Books.
Rose, H., & Rose, S. (Eds.). (1976). *Ideology of/in the natural sciences*. Cambridge, MA: Schenkman Publishing Company.
Rosser, S. (1985). A feminist perspective on science: Is reconceptualization possible? *International Journal of Women's Studies*, in press.
Rossiter, M. W. (1982). *Women scientists in America: Struggles and strategies to 1940*. Baltimore, MD: Johns Hopkins University Press.
Sayre, A. (1975). *Rosalind Franklin and DNA: A vivid view of what it is like to be a gifted woman in an especially male profession*. New York: W. W. Norton.
Sherman, J. (1980). Mathematics, spatial visualization, and related factors: Changes in girls and boys, grades 8–11. *Journal of Educational Psychology, 72*, 476–482.
Smith, R. J. (1985a). Star wars grants attract universities. *Science, 228*, 304.
Smith, R. J. (1985b). Academic consortia receive first star wars grants. *Science, 228*, 696–697.
Smith, R. J. (1985c). Scientific fraud probed at AAAS meeting. *Science, 228*, 1292–1293.
Spemann, H. (1938). Embryonic development and induction. New York: Yale University Press.
Spemann, H., & Mangold, H. (1924). Induction of embryonic primordia by implantation of organizers from a different species. Reprinted in B. H. Willier & J. M.

Oppenheimer (Eds.), *Foundations of experimental embryology*. New York: Hafner Publishing Company, 1974.

Tiger, L. (1977). The possible biological origins of sexual discrimination. In D. Brothwell (Ed.), *Biosocial man*. London: The Eugenics Society.

Vetter, B. (1980). Sex discrimination in the halls of science. *Chemical and Engineering News*, March, 37–38.

Wasser, S. K. (Ed.). (1983). *Social behavior of female vertebrates*. New York: Academic Press.

Watson, J. D. (1968). *The double helix*. New York: Atheneum Publishers.

Weisstein, N. (1979). Adventures of a woman in science. In R. Hubbard, M. S. Henifin, & B. Fried (Eds.), *Women look at biology looking at women*. Cambridge, MA: Schenkman Publishing Company.

Chapter 3

Critiques of Modern Science: The Relationship of Feminism to Other Radical Epistemologies

Elizabeth Fee

In the early stages of development of the women's movement, science and technology remained at the periphery of our vision. Consciousness raising and the construction of theory began with women's immediate experience and extended in time and space to history and anthropology; the imagining and reimagining of the world tended to be approached through art, literature, and psychology rather than through the technological imagination. At the core of the women's movement were the politics of personal relationships, sexuality, and reproduction; analysis began with the previously "private" sphere and moved onto the public stage. Science and technology, located in the public (male) world seemed, at first, to have little to do with the new politics of feminism.

This situation has changed dramatically within the last 5 years. In the politics of reproduction, the boundary disputes between technology and "nature" now appear to be erupting into a full-scale war. Very few feminists have followed Shulamith Firestone's (1971) original path of imagining liberation as a consequence of the developing technologies of reproduction. Most radical feminists have been critical, even hostile, toward existing and/or potential technologies, viewing them as new versions of men's control over women's bodies (e.g., Seaman, 1971). Now, within only a few years, the technological possibilities and realities have moved far beyond the level of these early debates; genetic engineering is rapidly transforming the meaning and nature of human reproduction, while feminist analysts race to catch up with the social and political implications of these changes for women's lives (Arditti, Klein, & Minden, 1984; Corea, 1985). The analysis of the production and application of scientific knowledge is moving toward the center of contemporary feminist theory and practice.

Genetic engineering is only one example of the potential for science and technology to transform women's lives on every level. New technologies of production are generating dramatic changes in the international/sexual division of labor: appropriate images of the new industrial worker include the young Asian woman working in an electronics plant in Taiwan, Manila, or Hong Kong, or the home worker coding sales orders on a computer network. Sectors of the economy traditionally reserved for women in a sex-divided labor market have been expanding while those reserved for men are declining; for those women confined to the home and family, decentralized home-based labor may break down the distinction between home and factory. Technologies that are changing the face of politics, communications, and warfare are also helping to change the relations between the sexes and the social meaning of gender.

We have been used to a virtual male monopoly of the production of scientific knowledge and discourses about science, its history and meaning. In response to the current possibility of transforming the social relations between the sexes, a conservative ideological movement within science has mobilized to defend inequality, protect the status quo, and create barriers to change: sociobiological theories are among the weapons of ideological warfare. Feminists, however, have not been silent; they have met these reactionary efforts with a positive explosion of new theoretical work, exploring the relationship between science, technology, and gender. This new body of work has encompassed several different and sometimes contradictory strands, but as a collective body of theory it seeks to defend women from ideological attacks conducted in the name of "science"—and also it goes much further in challenging the entire tradition of Western science by exposing how the foundations of our knowledge have been built on the assumptions of male domination and patriarchal power (Bleier, 1984; Harding & Hintikka, 1983; Hubbard, Henefin, & Fried, 1982; Keller, 1983; Keller, 1985; Lowe & Hubbard, 1983; Merchant, 1980).

This round of confrontation with scientific knowledge began as critiques of specific theories, historical and contemporary, and showed how these theories represented (often outrageously) biased accounts of women and of sex differences (e.g., Bleier, 1984; Haraway, 1978, 1979; Jordanova, 1980; Tanner, 1981). Critiques of specific theories were often initiated with the intent of improving science by removing some of its worst abuses, but they also tended to raise questions about science in general—particularly, the authoritarian epistemological claims made on behalf of scientific knowledge and method. A distinction may be made here between the humility of at least some practicing scientists, conscious of the narrowness and uncertainty of existing knowledge, and the arrogance of publicists and ideologists of science. The dominant culture ignores the uncertainty of scientific knowledge and presents an authoritarian view of scientific truth. Since the sciences are opaque to nonspecialists, these claims of expertise tend to be convincing and intimidating. And most women, even more than men, have been convinced that they are incapable of understanding science and mathematics. The sciences are presented as quintessentially male forms of knowledge: abstract, depersonalized, objective, authoritative. The woman who engages in scientific knowledge, who struggles and debates the meaning of its claims and pronouncements, is confronting power, breaking the taboo, eating the apple in the garden.

The exclusion of women from the production of scientific knowledge has been neither recent nor accidental. As Merchant, Keller, and others argued, the language and metaphors of the scientific revolution were clear: sexuality was the metaphor for the mediation between mind and nature. Mind was male, Nature was female, and knowledge was created as an act of aggression—a passive nature had to be interrogated, unclothed, penetrated, and compelled by man to reveal her secrets (Keller, 1985; Merchant, 1980). The identification of man as the knowing mind and woman as his connection to nature has been a continuing theme in Western culture. Women and human reproduction are consigned to the sphere of the "natural," while other human activities (and men) are assigned to the sphere of the "social." This language of women as especially "natural" beings is perhaps most explicit in 19th century evolutionary biology and anthropology, but the same assumptions appear in a multitude of literatures and languages. Simone de Beauvoir (1953) speaks, for example, of women immured in natural processes, held back from human transcendence and from the creation of history by the requirements of mere biological life. Woman is thus said to be enslaved to the species, cursed by her reproductive capacity. The immanent body is opposed to the transcendant mind; the creation of other human beings is seen as a passive act in contrast to the active creativity of "man" making himself (O'Brien, 1981). Woman's reproductive capacity is seen as an immense biological burden, condemning her to the world of nature, of the body, of emotions, and subjectivity. Again and again, we find the familiar dualisms of mind and body, culture and nature, rationality and emotionality, activity and passivity, objectivity and subjectivity, male and female; whenever men and women are contrasted, the values associated with men are those of transcendence and power.

The values thus associated with men are also those associated with science, or rather, they are the values and attributes of the scientist as the subject of knowledge. The object of knowledge, nature, is female. As Carolyn Merchant (1980), Donna Haraway (1979), Ludi Jordanova (1980), Evelyn Keller (1985), and others have shown, scientific enterprise, since at least the 16th century, has repeatedly been described and discussed in the language of sexuality and gender; science and medicine have been associated with sexual metaphors clearly designating nature as a woman to be unveiled, unclothed, and penetrated by masculine science. While the stereotypes of males and female become reified through the study of nature, the images and self-conception of the scientists become ever more firmly identified as masculine.

Lest anyone think that the sexual language of science was a historical accident of the 17th or 18th century, Sharon Traweek's (1982) fascinating anthropological studies of high energy physics laboratories suggest that the language of modern science continues to be sexualized. Richard Feynman, in his Nobel lecture, provides an eloquent example:

> That was the beginning, and the idea seemed so obvious to me and so elegant that I fell deeply in love with it. And, like falling in love with a woman, it is only possible if you do not know much about her, so that you do not see her faults. The faults will become apparent later, but after the love is strong enough to hold you to her. So, I was held to this theory, in spite of all difficulties, by my youthful enthusiasm. . . . So what happened to the old theory that I fell in love with as a youth? Well, I would say it's become an old lady, who has very little left that's attractive in her, and the young today will not have their hearts pound when they look at her anymore. But, we can say the best we can for any old woman, that she has become a very good mother and has given birth to some very good children. (Feynman, 1966, pp. 700, 708)

While the relationship of the scientist to nature was sexualized, physicists described their own relationships to each other in terms of physical images, such as the elastic collisions between billiard balls or, more positively, as the exchange of photons between interacting particles (Traweek, 1982). Such images reinforce the scientists' self-image as detached, objective observers: detached from each other, from the surrounding culture, from their own emotions and feelings, and from the possible applications and consequences of the knowledge they create. This highly prized detachment was discussed and applauded in the language of "degrees of freedom."

Because science has been so firmly identified as male, women in scientific fields have had to mediate between two worlds and a dual identity: to be a "real woman" is to be nonscientific; to be a "real scientist" is to be nonfeminine. Alice Hamilton, the first woman appointed professor at Harvard, expressed the choice as it appeared to many of her generation: marriage and a scientific career were simply incompatible and a professional woman had to choose (Sicherman, 1984). For our generation, the cultural accounts are more optimistic (although the employment opportunities may be little better). Women's magazines have produced a veritable industry of articles on the possibility or impossibility, tensions and stresses, of trying to "have it all": how to mediate the contradictory demands of professional, especially scientific, work and traditionally female responsibilities for family and human relationships. Should a woman scientist expect to cultivate a split personality, should she develop a gender-neutral persona for the laboratory, should she integrate her scientific work with a specifically female identity: how can she avoid being discounted as either too weak or too aggressive? She rarely makes a single choice, but rather a succession of choices, involving a whole series of practical decisions, consciously or unconsciously made. In this process, women internalize the cultural contradictions of gender in a constant, ongoing process of mediating opposing cultural demands.

Until recently, women scientists tended to deny that these choices were fundamental or problematic: while women were trying to gain access to scientific careers, it seemed strategic to deny that gender could be relevant to the production of knowledge. After all, the implicit understanding had always been that if there was any problem, it was women's problem – the organization of science itself was necessary to the production of knowledge and, therefore, beyond reproach. Women knew the rules and would abide by them, tailoring their thoughts and ideas to the dominant style of the laboratory in much the same way that business executives donned the suits deemed appropriate for the corporation. Any possible conflicts should be discussed in private rather than aired in public. A woman must look and work like a man, since any differences would automatically be defined as disabilities.

Something, however, happened to explode the discreet silence. Many of the discussions are still personal and private, but as women gain some real collective access to scientific fields, a range of questions has begun to enter the public domain. On a popular level, *Mademoiselle* published an article by a female medical student comparing medical training to boot camp, marked by a succession of tests of "toughness" such as willingness to kill a dog during the course of a set of experiments on the heart and circulatory system. Daniel Segal (1984), in his examination of the FYBIGMI (Fuck You, Brother, I Got My Internship) performance at an American medical school, went much further in detailing the alternate cycles of intimidation and humiliation, followed by equally intense congratulation and celebration, that characterize medical education and the socialization of physicians. He showed how alienation and callousness towards patients, racism and sexism, and casual jokes about sexuality and mortality were used to demonstrate professional toughness and to celebrate professional initiation. Sharon Traweek (forthcoming), in discussing the culture and socialization of the particle physics community, elaborates a complex route of professional initiation that is clearly incompatible with anything remotely resembling a normal female life cycle; the surprise is not that there should be so few women in high energy physics but that there should be any at all.

When feminist writers say there is something wrong with all this, they are to a certain extent reinforcing the old stereotypes and dualisms in concluding that contemporary science *is* in some fundamental ways a "masculine" endeavor. Instead of deciding that, therefore, something is wrong with women, however, they decided that something is wrong with science and, indeed, something is wrong with the way men relate to the natural world and to each other as well. The alternative to changing women is to change science. This subversive possibility implies the creation of new ways of experiencing/thinking about the relationship of knower to known and a new approach to the "natural," structured as a conversation rather than a command. Many feminist theorists have argued that women have different conceptions of the basic constituents of reality, different assumptions about their own relationships to the natural world, different views of the importance and connectedness of other people, more ready access to their own emotions and feelings, and distinct ways of assessing moral responsibility, based on a context of human relationships rather than abstract rights of isolated individuals (Gilligan, 1982). In other words, whether consciously articulated or not, women carry the seeds of an alternative ontology, epistemology, and ethics. While Western culture legitimizes only masculine ontologies, epistemologies, and ethics, the alternative epistemological assumptions making up women's world view remain subversive possibilities.

In the main, the literature on women and science has tried to do two things: to reveal the pro-masculine bias within existing science, in terms of its epistemological assumptions, methods, theories, "facts," and interpretations; and to argue for alternative subversive assumptions, methods, etc., of a different, feminist science. These discussions have framed the epistemological assumptions or preconditions of a feminist science: one in which no rigid boundary separates the subject of knowledge (the knower) and the natural object of that knowledge; where the subject/object split is not used to legitimize the domination of nature; where nature itself is conceptualized as active rather than passive, a dynamic and complex totality requiring human cooperation and understanding rather than a dead mechanism, requiring only manipulation and control. In such feminist imaginings, the scientist is not seen as an impersonal authority standing outside and above nature and human concerns, but simply a person whose thoughts and feelings, logical capacities, and intuitions are all relevant and involved in the process of discovery. Such scientists would actively seek ways of negotiating the distances now established between knowledge and its uses, between thought and feeling, between objectivity and subjectivity, between expert and nonexpert, and would seek to use knowledge as a tool of liberation rather than of domination (Fee, 1981). The basic claim of these discussions is that an alternative conception of science and nature is possible, but that its achievement is impeded by the rigid dichotomies and associated intellectual barriers inherited from the genderization of the world into two mutually exclusive realms. The ability to see beyond these dichotomies is not, of course, a biological capacity exclusive to women but is available to members of both sexes; the point is simply that women have more to gain by throwing off the chains of a gendered world that has crippled their movements so severely.

Evelyn Fox Keller (1983) brought us, in Barbara McClintock, a woman whose scientific work reveals many of the themes previously stressed by those of us talking about a feminist epistemology. McClintock views nature not as a passive, mechanical object ruled by externally imposed law, but as alive, growing, internally ordered, and resourceful. McClintock's attitude towards this "object," nature, gives new meaning to the relationship of objectivity and subjectivity and explodes them as rigid dichotomies. Had it not been for her extraordinary scientific accomplishments, McClintock might be dismissed as a romantic with a poetic involvement in the natural world, a woman incapable of maintaining a proper distance from the object of her study. Instead, she discovered a different approach to genetics, one that recognizes and permits the complexity of interacting systems, including the interrelationships of observer to observed, cell to organism, and organism to environment: an alternative to the dominance of a simple mechanical order or the dictates of "master" molecules within the cell.

McClintock's work thus provides us with an example of how highly elaborated scientific work can be based on epistemological assumptions opposed to those of the dominant, mechanical philosophy. Feminists, in turn, have claimed these epistemological assumptions as part of the framework of a new science: one that would transcend the gendered dichotomies of an older, more rigid and mechanical world. Intrinsic to this claim is the idea that women have some kind of privileged access to the new epistemology, not because of a biological difference but because of their different experience. Women's social experience leads to (or makes possible) an alternative view of the world and of nature. The most elaborated and most popular set of theories supporting the assertion that women see the world differently from the way men do are the post-Freudian psychoanalytic object relations theories variously presented by Nancy Chodorow (1978), Dorothy Dinnerstein (1976), Jane Flax (1983), Evelyn Keller (1985), and others.

In this paper, however, I argue that psychoanalytic theories are not an adequate basis for arguing the claims of a feminist science. I approach this question by looking briefly at several epistemological approaches suggesting that central themes of the feminist literature are echoed in alternate cultural accounts of science, as well as by radical and Marxist literatures. The existence of alternate epistemological accounts suggests that the central issue in criticizing science is one of power. Gender is but one of a series of dominant/dominated relationships whose intersections must be analysed and placed at the center of a feminist politics.

Let us look, then, at post-Freudian psychoanalytic object relations theory, as it is the main body of theory used to support and articulate the claim that we women see ourselves and the world around us differently from the way men do. Following Chodorow's highly influential discussion of the reproduction of mothering, object relations theory has occupied a central place in the theoretical arsenal of feminism. To summarize briefly, the theory states that little boys and girls growing up form different kinds of ego boundaries and have different experiences of their relationships to other people and to the external world. In the context of female mothering, little boys must form their gender identity by cutting themselves off from the mother, the primary love object, while little girls can continue to identify with the mother and do not experience the same abrupt break. In forming a masculine gender identity, little boys must deny and repress their early identification with the mother. The consequences of these early emotional patterns are far-reaching. In Chodorow's words

> Feminine personality comes to be based less on repression of inner objects, and firm and fixed splits in the ego, and more on retention and continuity of external relationships. From the retention of preoedipal attachments to the mother, growing girls come to define and experience themselves as continuous with others; their experience of self contains more flexible or permeable ego boundaries. Boys come to define themselves as more separate and distinct, with a greater sense of rigid ego boundaries and differentiation. The basic feminine sense of self is connected to the world, the basic masculine sense of self is separate. (Chodorow, 1978, p. 169)

A shortcoming of object relations theory is that it may appear to explain too much. The theory is based on modern, Western, nuclear families within capitalist economies—and it essentially concerns families in which an isolated full-time mother takes direct responsibility for childcare and housework, while the absent father is occupied in the labor market. The tendency, however, is to suppose that the sex-gender system with which we are most familiar remains, at least in essentials, always and everywhere the same. (Is not male dominance universal?) The problem is exacerbated by the fact that we know almost nothing about historical changes in parenting practices, about the effects of class on the psychodynamics of the family, or about the impact of women's work, single parenthood, or indeed, any number of other contemporary forms of family structure. Because of these limitations, some of those who based the idea of a feminist epistemology on psychoanalytic theory have concluded that the structure of the family in which mothering is left to women is the source of the problem; shared parenting between men and women would, it is sometimes claimed, eliminate the difference between male and female psychology, transforming the social organization of gender and eliminating sexual inequality. Such shared parenting should, by this account, transform our relationship to the natural world as well as to each other—including, presumably, our ways of creating science and technology. This approach places too much weight on the structure of family relationships as the cause, and eventually the solution, of all other social problems. Object relations theory cannot bear the full weight of explaining the larger historical structures of economic and political power. As Roger Gottlieb (1984) and others have argued, mothering as presently constituted is an integral and important part of the sex/gender system but is not a primary cause of that system. Exclusive female mothering can only reproduce political, cultural, and economic devaluation of women in a society *already* based on patriarchal power relations. This is not to say that female mothering is "merely" a consequence of larger structures of political and economic power: Chodorow's theory provides an essential part of a larger structural feminist account of power in showing how all the social relations of domination can become reproduced in the family, in individual psychology, and intimate human relationships.

To summarize the critique of object relations theory, I can state that any psychoanalytic theory can only be a theory of the reproduction of power relations; to find the sources of unequal power and domination we must look beyond the family. The theory of gender relations constructed within the family must be articulated with the social power relations of the larger society. One way of beginning this process in relation to the feminist critique of science is to examine some alternate epistemologies produced in reaction to other dominant/dominated relations of power.

Let us now briefly look at the relationship between feminist epistemologies of science developed within Western capitalist countries and epistemologies of science that presented as representing African, Indian, Chinese, and working-class perspectives on nature and natural knowledge. All of them appear to embody very similar ideas: conceptions of nature that in one context are denounced as masculine are in another denounced as European, colonial, white, or bourgeois.

Sandra Harding (in press), in a paper on "The Curious Coincidence of African and Feminine World Views," makes part of this case by counterposing sets of quotations from Vernon Dixon's writings on African philosophy with those of feminist writers. Dixon (1976) argues that the sense of self in the African world view is intrinsically connected with both community and nature; the individual is defined by his or her relationships within the community, and not, as in the West, by contrast to the collectivity. Just as the individual can exist only in relationship to the community, so too, can he or she exist only in relationship to nature: human culture exists as a part of nature, not in opposition to nature. The epistemological consequences of this perspective are profound. As Harding summarizes the argument

> Because the self is continuous with nature rather than set over and against it, the need to dominate nature as an impersonal object is replaced by the need to cooperate in nature's own projects. Coming to know is a process which involves concrete interactions, ones which acknowledge the role that emotions, feelings and values play in gaining knowledge, and which recognizes the world-to-be-known as having its own values and projects. (Harding, forthcoming)

Similarly, Russell Means (1980), in his speech "Fighting Words on the Future of the Earth," denounced all forms of "European" thought as devoid of a spiritual appreciation of the natural world and, therefore, as leading merely to different forms of exploitation of the earth and its natural resources. In contrast to what he terms the *European forms of domination*, Means emphasizes American Indian traditions of respect for nature as active and alive rather than passive or dead, and he celebrates ways of knowing that are conceptualized as an interrelationship with natural forces rather than a process of domination and subordination. The argument that only the "Western" or "European" approach has been successful in exploiting the earth's resources is directly challenged: the Indian approaches to nature allowed communities to live in their environments, both using and conserving natural resources, and thus permitting the reproduction of societies for thousands of years. By contrast, the European approach has proved capable of denuding the land or poisoning the waters within remarkably short periods of time; decades or even years of exploitation can do irreversible damage to the land, making it unfit to support human life. From Means' point of view, the description of such an approach as "successful" requires a dangerous perversity of definition and human purpose.

Consider, also, some of Joseph Needham's characterizations of Chinese science (Ronan & Needham, 1978). Chinese thought, says Needham, was profoundly non-Cartesian, refusing to make any sharp dichotomy between spirit and matter or between mind and body. Similarly, Chinese traditional medicine integrated spirit, mind, and body, diet and dreams, energy flows and physical sensations, and remains highly successful at an empirical level, while resisting all efforts to define it within the categories of physiological reductionism.

Chinese physics also remained impervious to mechanical materialism, atomism, and physical reductionism, remaining "perennially faithful to their prototypic wave theory of Yin and Yang . . . the Yin and Yang, and the Five Elements, never lent themselves to reductionism because they were always inextricably together in the continuum . . . never separated out, isolated or 'purified,' even in theory." Abstract knowledge of the either A or not-A variety was avoided in Chinese science in favor of nonexclusive relationships of forces. In general, "man's" position relative to nature was conceived as one of intimate and harmonious relationship rather than one of domination.

> The universe did not exist specifically to satisfy humans. Their role in the universe was to assist in the transforming and nourishing process of heaven and earth, and this is why it was so often said that humanity formed a triad with heaven and earth. It was not for man to question Heaven nor compete with it, but rather to fall in with it while satisfying his basic necessities. . . . Thus, for the Chinese, the natural world was not something hostile or evil, which had to be perpetually subdued by will-power and brute force, but something much more like the greatest of all living organisms, the governing principles of which had to be understood so that life could be lived in harmony with it. (Needham, 1976)

These are a few voices from the chorus of epistemologies of science from precapitalist societies. All argue for a more integrated understanding of "man's" role in the natural world. Some Marxist authors have argued that scientific abstraction and objectivity were the creations of capitalist development, specifically, the outcome of commodity exchange and the division of mental and manual labor. For example, Alfred Sohn-Rethel says

> The pattern of movement inherent in the exchange abstraction introduces then a definitive concept of nature as material object world, a world from which man, as the subject of social activities, has withdrawn himself. (Sohn-Rethel, 1978, p. 56)

This position seems to offer the theoretical possibility of relating the Marxist and feminist perspectives on scientific knowledge by grounding them together within history (Hartsock, 1983; Rose, 1983). Just as there are many forms of feminism, so too are there many different forms of Marxism. One tradition within Marxism has uncritically embraced scientific and technical knowledge as progressive. Lenin, for example, saw science as a neutral instrument that could serve the interests of whatever class controlled it (Claudin-Urondo, 1977). Other Marxist authors have taken a more critical attitude towards science and present themes (in a different language) also present in the feminist literature (Lewontin, Rose, & Kamin, 1984; Notworthy & Rose 1979; Rose & Rose, 1976; Schmidt, 1971). Raymond Williams, for example, says

> Out of the ways in which we have interacted with the physical world we have made not only human nature and an altered natural order; we have also made societies. It is very significant that most of the terms we have used in this relationship—the conquest of nature, the domination of nature, the exploitation of nature—are derived from real human practices: relations between men and men. . . . Capitalism, of course, has relied on the terms of domination and exploitation; imperialism in conquest has similarly seen both men and physical products as raw material. But it is a measure of how far we have to go that socialists also still talk of the conquest of nature, which in any real terms will always include the conquest, the domination of some men by others. If we alienate the living processes of which we are a part, we end, though unequally, by alienating ourselves. (Williams, 1980)

What, then, do we do with all this? Why do the themes presented in the feminist literature sound so similar to those of African, Indian, Chinese, and working-class or Marxist perspectives on philosophy and science? Clearly, this whole topic deserves more extended exploration but, as a first statement, we can say that it is not sufficient to describe the conception of science being criticized simply as male nor to make the structure of the family and psychoanalytic theory bear the full burden of causal explanation. Indeed, the conception of science defined as male in much of the feminist literature belongs to a specific period of the last 300–400 years and is characteristic of the period of capitalist development; it *is* European and also male, and white, and bourgeois: it is *also* a historical creation with boundaries in time (Lewontin et al., 1984). At present, we have at least three largely separate literatures, all critically addressing the forms of this science. In the literature of black and Indian liberation, it is addressed as white or European science; in the literature of feminism, it is addressed as male science; in at least some of the literature of Marxism, it is addressed as bourgeois science (Lecourt, 1976; Navarro, 1981).

Clearly, these different critiques need to be brought together; we need to understand the relationship between them and to explore the question, Is this the *same* critique, simply expressed in different forms, or are we dealing with three (or more) different sets of problems? It seems to me that any one of these critiques provides a partial, but incomplete, perspective—and each adds important elements otherwise missing in the analysis.

Each of these critiques addresses one set of dominant/dominated power relations articulated and reproduced within scientific knowledge, reflecting the unequal power relations in the social world; the critiques thus undermine the scientific legitimation of those dominant/dominated relations. In this view, scientific knowledge is a synthesis and reflection of dominant/dominated relations in the "natural" (human) world. This is not to say that the sciences do not *also* provide ways of understanding, using, manipulating, and controlling the natural world; the point is, they do so by means of our human relationships. And these relationships are unequal in terms of power across several boundaries: class and race, as well as of gender. By necessity, all of these power relations are reproduced within scientific knowledge; the scientist, the creator of knowledge, cannot step outside his or her social persona and cannot evade the fact that he or she occupies a particular historical moment. The idea of a pure knowing mind outside history is simply an epistemological conceit. Reflected within science is the particular moment of struggle of social classes, races, and genders found in the real, natural, and human world.

Power, then, cannot simply be discussed in terms of male domination; maleness is articulated within a set of power relations of race, class, and nationality. Similarly, femaleness is not a single thing but is also articulated within a set of power relations—again, class, race, and nation. These forms of domination are not separate, or exterior to each other, but are integrated. Thus, you cannot be a woman without being a woman of a certain class, race, and country; similarly, you cannot be a woman without being a woman at a particular moment in history—and that moment in history will carry its own definition of what it means to be a "woman," what it means to belong to a certain class, race, and so on. All of these terms are continually being redefined in the context of ongoing political and ideological struggles; they are never static.

What this means for feminism is that, in one sense at least, there are many feminisms – for feminism must be (either explicitly or implicitly) articulated with class, race, and national struggles. There is no feminist position that can transcend the boundaries of class and race, nor should this be the aim – any more than it is possible, for example, to take a class position that ignores the divisions of gender. We can speak of gender, class, and race as simple abstract categories, but analysis of any specific historical–political conjuncture requires that they be understood in their relation to one another. Each form of analysis has something to offer the others: feminism takes the power relations of the public social world and shows how these are reproduced within the structures of the family, sexuality, and personal life. This analysis must at the same time be related back to the larger structures of reproduction of social and economic power. Feminism and feminist analysis are an essential part of human emancipation, but they are not the whole of the struggle of human liberation.

What this means for science is that the nature of our knowledge will continue to develop in response to these ongoing struggles – not simply in terms of the male–female, dominant/dominated relation but also with respect to class, race, and national struggles. There is not, therefore, a single feminist science that represents *only* the interests of women as a unified group. There *is*, however, a feminist perspective on science that shows the ways in which gender-based dominance relations have been programmed into the production, scope, and structure of natural knowledge, distorting the content, meaning, and uses of that knowledge. Feminist perspectives on science should be helpful in understanding other relations of dominance – as feminism, by extension, should question all forms of domination. Each of the specific struggles with and within scientific knowledge and practice should, if they are not too narrowly conceived, aid the other forces struggling for human liberation.

Acknowledgements – I would especially like to thank Kay Carter, Ruth Bleier, Ruth Finkelstein and Jane Sewell for their friendship and support, and Ruth Finkelstein for her invaluable assistance in revising and editing this essay during several hot and trying days this summer.

REFERENCES

Arditti, R., Klein, R. D., & Minden, S. (1984). *Test-tube women: What future for motherhood*? New York: Routledge and Kegan.

Bleier, R. (1984). *Science and gender: A critique of biology and its theories on women*. Elmsford, NY: Pergamon Press.

Chodorow, N. (1978). *The reproduction of mothering: Psychoanalysis and the sociology of gender*. Berkeley and Los Angeles: University of California Press.

Claudin-Urondo, C. (1977). *Lenin and the cultural revolution*. Sussex, England: Harvester Press.

Corea, G. (1985). *The mother machine: Reproductive technologies from artificial insemination to artificial wombs*. New York: Harper and Row.

de Beauvoir, S. (1953). *The second sex*. Translated and edited by H. M. Parshley. New York: Alfred A. Knopf.

Dinnerstein, D. (1976). *The mermaid and the minotaur: Sexual arrangements and the human malaise*. New York: Harper and Row.

Dixon, V. (1976). World views and research methodology. In L. M. King, V. Dixon, & W. W. Nobles (Eds.), *African philosophy: Assumptions and paradigms for research on Black persons*. Los Angeles: Fanon Center Publications, Charles R. Drew Postgraduate Medical School.

Fee, E. (1981). Is feminism a threat to scientific objectivity? *International Journal of Women's Studies, 4*, 378–392.

Feynman, R. P. (1966). The development of the space-time view of quantum electrodynamics. *Science, 153*, 699–708.

Firestone, S. (1971). *The dialectic of sex: The case for feminist revolution*. New York: Bantam Books.

Flax, J. (1983). Political philosophy and the patriarchal unconscious: A psychoanalytic perspective on epistemology and metaphysics. In S. Harding and M. B. Hintikka (Eds.), *Discovering reality: feminist perspectives on epistemology, metaphysics, methodology, and philosophy of science*, pp. 245–281. Dordrecht, Boston, and London: D. Reidel Publishing Company.

Gilligan, C. (1982). *In a different voice: Psychological theory and women's development*. Cambridge, MA, and London: Harvard University Press.

Gottlieb, R. G. (1984). Mothering and the reproduction of power: Chodorow, Dinnerstein, and social theory. *Socialist Review, 14*, 93–119.

Haraway, D. (1978). Animal sociology and a natural economy of the body politic. *Signs, 4*, 21–60.

Haraway, D. (1979). The biological enterprise: Sex, mind and profit from human engineering to sociobiology. *Radical History Review, 20*, 206–237.

Harding, S. (in press). The curious coincidence of African and feminine world views. In D. Meyers and E. Kittay (Eds.), *Women and moral theory*. Totowa, NJ: Rowman and Allenheld.

Harding, S., and Hintikka, M. (Eds.). (1983). *Discovering reality: Feminist perspectives on epistemology, metaphysics, methodology and philosophy of science*. Dordrecht, Boston, and London: Reidel.

Hartsock, N. (1983). *Money, sex and power*. New York and London: Longmans.

Hubbard, R., Henefin, M. S., & Fried, B., (Eds.). (1982). *Biological woman: The convenient myth*. Cambridge, MA: Schenkman.

Jordanova, L. (1980). Natural facts: A historical perspective on science and sexuality. In C. MacCormack and M. Strathern (Eds.), *Nature, culture and gender*, pp. 42–69. Cambridge, England: Cambridge University Press.

Keller, E. F. (1983). *A feeling for the organism: The life and work of Barbara McClintock*. New York and San Francisco: W. H. Freeman.

Keller, E. F. (1985). *Reflections on gender and science*. New Haven and London: Yale University Press.

Lecourt, D. (1976). *Proletarian science: The case of Lysenko*. London: New Left Books.

Lewontin, R. C., Rose, S., & Kamin, L. J. (1984). *Not in our genes: Biology, ideology, and human nature*. New York: Pantheon.

Lowe, M., & Hubbard, R. (Eds.). (1983). *Woman's nature: Rationalizations of inequality*. Elmsford, NY: Pergamon Press.

Means, R. (1980, December). Fighting words on the future of the earth. *Mother Jones*, pp. 22–38.

Merchant, C. (1980). *The death of nature: Women, ecology and the scientific revolution*. New York: Harper and Row.
Navarro, V. (1981). Work, ideology and science: The case of medicine. *Working Papers on Marxism and Science, 1*, 23–55.
Navarro, V. (1983). Radicalism, Marxism and medicine. *International Journal of Health Services, 13*, 179–202.
Needham, J. (1976). History and human values: A Chinese perspective for world science and technology. In H. Rose & S. Rose (Eds.), *Ideology of/in the natural sciences*, pp. 255–256. Cambridge, MA: Schenkman.
Notworthy, H., & Rose, H. (1979). *Counter-movements in the sciences*. Dordrecht, Boston, and London: Reidel.
O'Brien, M. (1981). *The politics of reproduction*. Boston and London: Routledge and Kegan Paul.
Ronan, C., & Needham, J. (1978). *The shorter science and civilization in China: 1*. Cambridge, England: Cambridge University Press.
Rose, H. (1983). Hand, brain and heart: A feminist epistemology for the natural sciences. *Signs, 9*, 73–90.
Rose, H., & Rose, S. (Eds.). (1976). *Ideology of/in the natural sciences*. Cambridge, MA: Schenkman.
Schmidt, A. (1971). *The concept of nature in Marx*. London: New Left Books.
Seaman, B. (1971). *The doctor's case against the pill*. New York: Avon.
Segal, D. (1984). Playing doctor, seriously: Graduation follies at an American medical school. *International Journal of Health Services, 14*, 379–396.
Sicherman, B. (1984). *Alice Hamilton: A life in letters*. Cambridge, MA: Harvard University Press.
Sohn-Rethel, A. (1978). *Intellectual and manual labour: A critique of epistemology*. Atlantic Highlands, NJ: Humanities Press.
Tanner, N. (1981). *On becoming human*. Cambridge and London: Cambridge University Press.
Traweek, S. (1982). *Gossip in Science*. Paper presented at the annual meeting of the American Anthropological Association, Washington, DC.
Traweek, S. (Forthcoming). *Taking space and making time: The culture of the particle physics community*.
Williams, R. (1980). *Problems in materialism and culture: Selected essays*. London: New Left Books.

Chapter 4

Beyond Masculinist Realities: A Feminist Epistemology for the Sciences

Hilary Rose

> I am convinced that "there are ways of thinking that we don't yet know about." I take those words to mean that many women are *even now* thinking in ways which traditional intellection denies, decries, or is unable to grasp. Thinking is an active, fluid, expanding process; intellection, "knowing" are recapitulations of past processes. (Rich, 1974, p. 284)

To ask, Is a feminist science possible? is to return to our own history of struggle and the contradictory relationship of feminism to science. For the second wave of feminism, science and technology have not—with the almost single and certainly exceptional voice of Shulamith Firestone—been seen as progressive for women's interests. There has been little chance of invoking the metaphor, unhappy or otherwise, of courtship and marriage that has been used to capture the spirit of the relationship between Marxism and feminism. Where the radical science movement of the 1960s had to free itself from the progressivist claims of science—to show that science was not even neutral but oppressive and antithetical to human liberation—women, already outside such progressivist claims as a result of their very exclusion from science, saw clearly that modern science and technology served as means of their domination and not their liberation.

Overtly relegated to nature by the recrudescence of the patriarchal determinism of sociobiology, feminists have learnt to uncover and contest the practices of a phallocentric science. In claiming a place in culture, feminism has had to think much more deeply about our social relationship with other human beings and our relationship to nature. No abstract materialism, the materialism of feminism in a direct and practical way finds itself defending nature, understanding that we, our bodies, ourselves, are part both of nature and of culture.

The recurrent mood, as and when the feminist movement preoccupied itself with science, has been one of anger. The anger extended from a sense of injustice at being shut out of an activity that some women, despite the engendered rules of the game, always wanted to take part in to an overwhelming sense of fury that masculinist science and technology are part of a culture of death. The ideology of science, proclaiming its objectivity, freedom from values, and dispassionate pursuit of truth, has excluded women and been integral to their cultural domination, has harmed women's bodies (in their best interests, of course), and threatens the environment itself.

Second wave feminism began relatively slowly to analyse and contest science, to see the connections between "it" (whatever it was) and those issues that the movement defined as its own. There were good reasons why the movement was slow; its central preoccupation was with women's shared experience, to reclaim what had been denied or trivialized out of existence and return it to social and political existence. The culture has changed in many ways since those early, path-breaking years of the late 1960s and early 1970s. Then, to consider housework, motherhood, sexuality, love, birth control, abortion, and male violence as central was to work against the grain of an arrogant sexism. Feminism began to understand the significance of body politics, so that the struggle for the repossession of our bodies, including knowledge about them, was to become central to feminist struggle. The very process of examining these everyday aspects of women's lives created new concepts, new names.

Naming is rightly seen within feminism as offering transformative powers. It brings into existence phenomena and experiences hitherto denied space in both nature and culture. Nor was this previous denial a matter of chance, for in the fierce resistance to the new names, the new concepts, we learn that they were not merely unacknowledged aspects of reality but were actively erased by the values of the dominant culture. The act of naming, above all when the words become part of the language of a significant social group seeking to take its place in society, simultaneously affirms a changed consciousness of reality and contests existing hegemony.

Feminism also found that its reasoning differed from the linear form of cognitive reasoning that took delight in dichotomies. Metaphors of spinning (Daly, 1979) and quilt-making (Balbo, 1983) are invoked as feminism speaks about its distinctive ways of thinking, feeling, and acting in the world. But, while I shall return to this issue of the sense of totality encompassed by feminism, here I only want to say that it offers us an insight into the fragmentary way that women's experiences of science, like everything else, were brought together, so that quite suddenly the pattern of the whole was transparent to everyone.

While feminism touched women's lives the world over and drew increasing numbers of women into its vortex, it is nonetheless true that the movement has been strongest – and it is here that the discussion of science is most advanced – within the old capitalist societies. This is not to say that feminists in actually existing socialist societies and third-world or black feminists within advanced industrial societies have experienced science and technology in a particularly favourable way; rather that, for strategic reasons, their attention has been primarily focussed elsewhere (e.g., Mass, 1976).

TAKING STOCK

While the specific question I seek to address in this paper concerns the possibility of a distinctly feminist epistemology, I shall begin by reflecting on the main achievements of the feminist critique of science and technology. This is partly because I want to honour those pioneering second wave voices that broke the long silence and because, through listening to them and taking stock of that rich and diverse scholarship, we can fashion a map to guide our present theoretical journeys. My feeling is that only because of this work are we able to recover and systematically pursue the insights of an earlier wave of feminism—above all Virginia Woolf's (1938) compelling aphorism, "Science it would seem is not sexless; she is a man, a father and infected too." Conscious of the price of classificatory schemes when a fresh reading of almost any piece suggests another and equally valid allocation, I want to suggest that feminist writing on science covers four main themes: why so few?; the recovery of Hypatia's sisters; the contestation of patriarchal science; and the feminist critique of epistemology. Yet, threaded within all four is a fifth theme, which offers us utopian alternatives, playing skillfully with science and technology to suggest new and pacific relationships between humanity and nature and among human beings themselves. In a situation where the dominant quality of men's science fiction writing has been not simply conservative but rather reactionary, in that it seeks not only to preserve but actively worsen existing social relationships—merely infusing them with ever-more sophisticated technology—feminist science fiction writing offers the possibility of a new, utopian society. From the delightful but essentially static utopia based on women's values of Charlotte Perkins Gillman's *Herland*, we are now invited to share the continuously contested and deliberately open-ended projects of Ursula Le Guin, Joanna Russ, and perhaps most of all, Marge Piercy. It is not by chance that feminists writing or talking about science and technology constantly return, as I do here, to these empowering alternative visions. Listen to Piercy's future person Luciente.

> Our technology did not develop in a straight line from yours. . . . We have limited resources. We plan co-operatively. We can afford to waste . . . nothing. You might say our—you'd say religion?—ideas make us see ourselves as partners with water, air, birds, fish, trees. (Piercy, 1976, p. 125)

In what follows, I first want to review the achievements of these main themes of feminist writing. On the basis of this review, I want to ask why women have been excluded from science and why science, therefore, is a peculiarly masculine institution. I seek to ground the answer to these questions within the theory of labour, as I see the case of science as only an instance of the general division of productive and reproductive labour between men and women. Because scientific knowledge flows from practice within the world, a feminist epistemology, rooted in the caring labour of women's work, will be qualitatively different from the one-sided materialism of masculinist science. Finally, I will conclude by asking whether such an epistemology will truly form a science that will succeed from the existing forms or whether, as in a post-modernist conception, a plurality of knowledges will coexist indefinitely.

WHY SO FEW?

Alice Rossi's (1965) question concerning the sparcity of women in the peculiarly masculine occupation of scientist, posed two decades ago, has been asked and answered in very different ways by feminists and by their enemies. While overtly patriarchal theorists claimed that this expression of the division of labour is sexual, and that "anatomy is destiny," feminists emphasized the social construction of gender. The distinctive contribution of feminists who have survived in what Anne Sayre (1975) rightly described as "an especially male profession" has been a sharp account of the male-operated exclusion mechanisms of science from physics to psychology (e.g., Couture-Cherki, 1976). Naomi Weisstein's paper is a classic in this genre. She begins

> I am an experimental psychologist, doing research in vision. The profession has for a long time considered this activity, on the part of one of my sex, to be an outrageous violation of the social order and against all the laws of nature. Yet at the time I entered graduate school in the early sixties I was unaware of this. I was remarkably naive. (Weisstein, 1977, p. 187)

Later, Weisstein was to add a footnote, observing that she had since realised how exceptional it was for a woman to have survived as far as graduate school. Other natural scientists were equally blunt. Diane Narek (1970) teaching physical sciences wrote, "The only reason that there aren't any more women scientists and technicians is because the men don't allow it" (p. 329).

Rita Arditti (1980) noting how common it was for women scientists to marry men scientists, often in the same field, saw that "All had secondary positions to their husbands regardless of ability; their loyalty as wives had led them to accept precarious work situations in which their research was dependent on their marriages" (p. 361). Arditti also reported on the hostility she experienced within the laboratory when she became involved with the women's liberation movement. From Evelyn Fox Keller (1977) we learned of the "sea of seats" and of the refusal of her male supervisor to believe that she could solve a mathematical problem unaided by a man. Applying for my first research post, I had a one-to-one interview during which I was sexually harassed; angry and humiliated, I withdrew my application. A demonstrably effective strategy in keeping the number of women researchers down, at least on a short-term basis.

What were the conditions through which any woman has survived and in some cases made important contributions to knowledge? The attempt to ask distinguished scientists to reflect on their lives has not been particularly successful. Two major collections were made, one by the New York Academy in the 1960s, one published by UNESCO (Richter, 1983). (Particularly ironic is that the male editor had never appointed a woman scientist to the laboratory he directed—but so it goes.) These autobiographical accounts are in themselves by and large unrevealing; the difficulties of personal life are ironed out; the emphasis is on the excitement of science. Yet, reading across these autobiographical accounts typically shows a highly privileged class origin and the unusual support and encouragement of a scientist father or husband.

I remember asking one of the very few women Fellows of the Royal Society, Dorothy Needham, why the biochemistry laboratory of Gowland Hopkins had served, in the interwar period, as a refuge not only for brilliant Jewish scientists fleeing Nazi Germany (e.g., Hans Krebs and Fritz Lipman), but also for so many brilliant women. These included Dorothy Wrinch, Dorothy Needham, Marjory Stephenson (who was to succeed Hopkins in the chair), and Barbara Holmes. Needham explained that it was through Hopkins' daughter, Barbara (Holmes), whose love of science he had actively encouraged. The class element was nonetheless important, for these women were not given any income; most of the work for which Needham herself received scientific recognition was done without a "proper job." One woman scientist of this generation described how she was paid exactly the cost of replacing her own labour in child care, not a way in which male scientists' salaries have been arranged.

Almost certainly, this particular generation of brilliant women scientists in England was possible only because science was, at that period, still something of a craft industry in which the daughters and wives of leading scientists enjoyed a certain privileged access to laboratories. As science became industrialised, these familial mechanisms were lost, so that access into the upper echelons of the scientific labour market became as problematic for women there as elsewhere. Science was relatively late to industrialise, and the access some women derived from their familial position has only recently been lost. Big science is thus more inhospitable to women than little science; its large and expensive toys are increasingly reserved for the boys. It is interesting that those areas where relatively low capital-to-labour ratios prevail still remain comparatively open to women, above all biology.

While most industrial countries have made some attempt to increase the percentage of women scientists and engineers – in Britain, for example, 1984 was the year of WISE (Women Into Science and Engineering) – few believe that reversing the present imbalance is an easy task (e.g., Kelly, 1981; Walkerdine, 1981). Indeed, despite some gains, such as the increase of women in medicine, losses are also evident; the proportion of women is lower in some areas compared with the interwar period. It is not simply difficult to get into science as a school subject, it is difficult to stay in. The spatial and time demands of being a laboratory scientist are much more in conflict with the demands of child care than being, say, a historian or a sociologist. While both the woman laboratory scientist and the woman historian may have in common the problem of the double day, the former has much less flexibility in choosing when or where to work. Sometimes, when considering the pressures against women entering and staying in the laboratories, so much the province of men, we may be forgiven if we pose Rossi's question more dourly and ask, Why any?

HYPATIA'S SISTERS RECOVERED?

The extraordinary difficulties confronting women who sought to enter science and the correspondingly extraordinary achievement of those who somehow managed to do so, made their erasure from the history of science an imperative matter for correction by feminist historians. Lynn Osen's (1974) study of women in mathematics is interesting here. Her biographical accounts of women mathematicians from Hypatia, herself, to Emmy Noether again emphasise the significance of sympathetic men members in their families providing encouragement and practical support to enable their talents and interests to develop. Several of the women mathematicians were members of mathematical families. In these, the dynastic tradition of the family members' as part of a scientific culture eroded the distinctive Victorian conception of the two spheres, the public and the private. Even well into the 20th century, individual scientists had laboratories in their own homes and carried out significant research from them, whether it was J. S. Haldane in England or Rita Levi-Montalcini in Italy (although granted the latter was under the special and very negative conditions of fascism). That such practices were possible indicates the craft state of the production of science and that particular women were thus aided in mediating between the public and the private.

While we need more historical work to map such a thesis – as industrialisation has entered each discipline at different historical periods – there are parallels with the history of other skilled occupations that sustain such a view. Certain occupations such as brewing and optics had some women masters, typically widows but sometimes daughters; they had picked up sufficient knowledge to take over in the event of a husband or father's death (Pinchbeck,

1930). The separation of home and factory and the systematic exclusion of women from apprenticeship, closed this possibility. It is not by chance that Ziman (1969) speaks of science as *public knowledge*. Mystified though this concept may be, it is part of the masculine occupation of the public sphere that women and women's knowledge is excluded from science. If we began with Rossi's question Why so few?, as the history of women's experiences in science begins to uncover, we are left marvelling at the strength of women who persisted and compelled recognition in this area. Sayre's passionate defence of Rosalind Franklin probably did more than any other single book to demonstrate the erasure of a particularly brilliant woman. That she challenged the accreditation system of science, in the particular area of DNA, that keystone of the biological revolution, was an act of profanity. No less than two (Watson and Wilkins) of the three male Nobel prizewinners felt it necessary to modify their accounts of the history of the discovery of the structure of DNA. Even more importantly, it enabled similar accreditation scandals to surface, such as the doubts about Hewish receiving a Nobel prize for work leading to the discovery of quasars largely carried out by his graduate student, Jocelyn Bell, or the questionable passing over of Candace Pert for the Lasker Award for work related to neurotransmitters. While there is little evidence that the accreditation system has become less loaded, there is greater realism about the extent of the bias built into the system. Equally, historians seem now to be more sensitive to gender issues in their own accounts of scientific developments: Hypatia's sisters have come into increasing visibility, if not into justice.

This sense of the particular nature of the scientist's knowlege itself makes Evelyn Keller's (1983) biography of Barbara McClintock so interesting. The Keller study is part of a second wave of biographical research, able to go beyond the painful mapping of exclusion or of compromise between private and public life and, instead, explores the specific contribution of a woman scientist whose life and work has been characterised by a high degree of personal autonomy and recognition. McClintock's awareness of her distinctive approach both to choice of research questions and methodology is made evident. The study of McClintock speaks of "a feeling for the organism," of "letting the material speak," and thus shares the conception of science as developing in a more complex way than the purely cognitive model most male science celebrates. Of course, Michael Polanyi's (1967) concept of tacit knowledge tries to discuss what is involved in creative scientific experimentation. By emphasising the mystical element in McClintock's approach, Keller points to her alternative philosophy of the relationship not only between the scientist and nature but the scientist's theory of organisation within nature. Thus, where cellular organisation was based on an assumption of a hierarchy in which the genetic material plays a leading role, McClintock's ideas turn on a more holistic, less hierarchic, conception of the organism.

CONTESTING PATRIARCHAL SCIENCE AND TECHNOLOGY

> It's that race between technology in the hands of those who control and insurgency—those who want to change the society in our direction. In your time the physical sciences had delivered weapons technology. But the crux we think is in the biological sciences. Control of genetics. Technology of brain control. Birth to death surveillance. Chemical control through psychoactive drugs and neurotransmitters. (Piercy, 1976, p. 223)

The 1970s saw a new wave of biological determinism committed to the renaturalisation of women; to an insistence that, if not anatomy then evolution, X chromosomes, and hormones were destiny; and to the "inevitability of patriarchy." Where the movement brought women from nature into culture, a host of greater or lesser sociobiologists joined eagerly in the effort to return them to whence they had come.

There was a twofold response to this attempt to use the ideology of science to reassert the status quo against the achievements of the movement. First, it changed perceptions within the women's liberation movement at large about the significance of science. Where hitherto the movement had tended to dismiss the activity of science as peripheral, the claims made by science now came to be seen as a threat to women's struggles. The consequence of this was to render the work of feminist scientists more significant to the movement at large. Among the feminist scientists themselves, the wave of biological determinism resulted in a conscious coming together to develop the scientific arguments against sociobiology's claims. The result was a host of pamphlets, conferences, symposia, and books: Bleier (1984) *Science and Gender*; Brighton Women and Science Group (1980) *Alice Through the Microscope*; Hubbard, Henifin, and Fried (1979) *Women Look at Biology Looking at Women*; Hubbard and Lowe (1979) *Genes and Gender*; Lowe and Hubbard (1983) *Woman's Nature*; Leibowitz (1978) *Females, Males and Families: A Biosocial Approach; Sayers (1982) Biological Politics*; Science for the People Collective (1977) *Sociobiology as a Social Weapon*; Tobach and Rossoff (1978) *Genes and Gender*.

Feminist anthropologists, ethologists, psychologists, and biologists showed that the claims to ground the gender division of labour in hormones, in evolutionary sociobiology, or in terms of just-so stories derived from ethological observations of other species were based on bad science—in the classical sense of the term—weak theory, inadequate and misinterpreted data, poor experiments, and inadmissable extrapolations between observations made on rats or ducks to humans. This cavalier approach to the limits of scientific method would not be acceptable in any less ideologically charged task than the legitimation of male domination and female subordination as rooted in biology and, therefore, natural.

The debate was and is waged in both the popular and the scientific domain. So-called scientific popularisation seized on and inflated sociobiological claims. Thus *Playboy* informed its readers that male promiscuity was a biological part of every man's birthright, whilst eventually feminist critique forced journals like *Animal Behaviour* to abandon its sexist language. It is now editorial policy that it is no longer acceptable to describe the sexual activities of mallard drakes as "rape." Managerial claims that "raging hormones" rendered women unfit for senior executive or political jobs were contrasted with the transformation in endocrinology resulting from the new feminist scholarship. The feminist critique of oppressive dichotomies within biological thinking of male/female, nature/culture, androgen/oestrogen, cognitive/affective, left brain/right brain has played a major part in revealing the ways in which gender has affected the development of science.

THE FEMINIST CRITIQUE OF EPISTEMOLOGY

The fourth area of the new scholarship sprang primarily from the work of feminist sociologists, historians, and philosophers of science. It began, in a sense, by reflecting on the masculinist ideology of science revealed in the debates about sociobiology and "woman's place" and showed how these were not new phenomena but were rooted in the history and tradition of a masculinist Western science. Whatever the fashionable biological measure of the moment, it appeared that women were deficient in it, from brain size to thigh length to body metabolism. The norm was male, women were by definition not simply different but inadequate, failing to reach the male norm. But the analysis was to become more wide-ranging, calling into question key aspects of the scientific tradition, such as its claims to a subject-free objectivity, to reveal a systematic relationship to nature premised on control and domination. At a theoretical level, the Frankfurt School had already begun to speak of the falsity of the fact/value split and had shown how the metaphor and practice of domination ran as a central thread through bourgeois and Marxist approaches to science (e.g., Leiss, 1972; Schmidt, 1971). This reevaluation and criticism of the scientific endeavour found resonances within the new social movements of the late 1960s, opposing the Vietnam war abroad and pollution everywhere. It was not entirely by chance that the scientific and popular contribution to the defence of the environment and humanity's place within it should have been made by a woman biologist and writer, Rachel Carson (1962). The initial dismissal of Carson and the ecology movement by the scientific community can be seen as an attempt to reassert traditional masculinist concepts of domination against the feminine value of harmony with nature, stressed by Carson. The masculinist emphasis on domination, against the need for a nonviolent relationship with nature, became more clearly visible both in the writings of the scientists themselves and as part of the project of the new social movements of the 1970s and 1980s.

The most sustained feminist development of the thesis of the domination of nature has been Caroline Merchant's (1980) account of the scientific revolution of the 17th and 18th centuries, *The Death of Nature*. She draws attention to the repeated metaphors of male domination, rape, and despoilation that characterise the writings of the father of the scientific revolution, Francis Bacon. She goes further to show how such metaphors ran systematically through the writings of scientists and philosophers of science since Greek and Roman times, albeit ignoring the critical tradition of Marxism, such as that developed by the Frankfurt School, which seeks a reconciliation between nature and culture, or that offered by Eastern scientific thought, particularly in its body politics. Nonetheless, these caveats apart, her argument is that the logical outcome of the masculinist theory and practice of the domination of nature is precisely the exterminism of today's science. Merchant counterposes this masculinist ideology with the harmonious relationship with nature sought by the twin and allied new social movements of ecology and feminism.

WOMEN'S WORK AND WOMEN'S KNOWLEDGE

Feminist scholarship then began by rendering visible the invisibility of women in science, it started to rescue Hypatia's sisters, restoring them from a manmade oblivion to history; it contested biological determinism's attempts at a renaturalisation of women and, in so doing, began to transform the theory and practice of science. And, it has endeavoured to show the ideological nature and consequences of a science that is not merely bourgeois but also masculinist. What is has not yet done is to show *why* women are excluded from science. To answer this question and to show, first, how this exclusion is rooted in material reality, and, second, what are the consequences for a feminist epistemology of science, requires an understanding of a materialist theory of labour.

Such a materialist theory requires that we understand the social function of excluding women from science: in whose interests does it occur and in what ways does it serve those interests? Only then can we effectively contest the exclusion and, insofar as all scientific knowledge flows from practice, from labour in the world, can we develop the characteristically feminist forms of knowledge that flow from women's work in the world. We have to begin by seeing scientific work as the particular example of a general division of knowledge between men and women in the world, which allocates characteristically different forms of work to men and to women. These different forms of work are found in both paid and unpaid labour, and are also reflected in an unequal division of labour time between men and women.

The analysis of the segregated labour market, nationally and internationally, shows that women are concentrated not only in low-paid work but frequently in work of a particular kind, human service work and menial work, a remarkable echo of women's work within the home. It is precisely in those societies where the largest proportion of women are employed in the paid labour market that we see the sharpest expression of segregated occupations and the greatest extent of part-time work. The latter, which is ideologically proposed as "choice" and as a gain for all, is for women in reality structured through the greedy time demands of the double day. Despite the ideology of science being above gender, this holds within the scientific labour market as much as any other.

In science therefore it is true, but only in a particular sense, that there are few women. In the United States, as in Britain, the academic staffs of science and engineering departments are almost all men, though the pattern varies from engineering (where there are almost no women faculty and only 8% women students) through physics and chemistry (which are slightly more mixed) to biology (where there is the greatest proportion of women as academic staff, although still a minority). But, while few women are in evidence in advanced science and technology education, that is not to say that none are to be seen at all. Women clean the floors, under the supervision of men supervisors; women act as technicians, under men senior technicians; they work as waitresses under men catering officers; and they work as secretaries typing letters dictated by men and generally smooth interpersonal relations. The point is—and it has to be made again and again—that woman's paid work, especially in the science or technology laboratory, echoes just what she does at home, except that at least there she is relatively free to get on with it at her own pace. The laboratory is simply part of the segregated labour market.

While anger fuelled the search for explanations of the unequal division of paid and unpaid labour in society, it also meant that the search is not merely for explanatory but for transformative theory. This requirement means that what was to become a passionate debate, particularly in Britain, over the social origins of domestic labour, was evaluated not only for its own internal theoretical consistency but also as to how it measured up to experience.

My overall feeling about the present state of the debate, which I shall make clear in my discussion in the following paragraphs of Nancy Hartsock's work and in the discussion of my own, is that a feminist materialism is almost certainly and necessarily dualist, recognising both the capitalist and patriarchal modes of production to which women's work and women's lives are meshed. But, I see this as a stepping-stone, corresponding both to the profound restructuring of the socioeconomic organisation of society and to the array of old and new social movements. Our theoretical inability to transcend dualism is, in a materialist analysis, connected to nothing less than our difficulty in actual practice at finding the alliances and new forms of political struggle between the old labour movement and the new social movements of ecology and feminism to bring about social revolution. Very much like Sylvia Walby (in press), I see the theoretical tension within dualism as not just a better tool for analysis but also a more creative framework for struggle.

Hartsock's "The Feminist Standpoint" (1983) represents an unequivocal response to this dualism. She uses Marx's materialist theory of knowledge to grasp what she unflinchingly labels the "sexual division of labour." She is well aware of the dangers of this strategy but argues, following Sara Ruddick (1980, p. 364), that some features of human life are "invariant and nearly unchangeable" while others are "certainly changeable" and points to the fact that women bear children (still) as evidence of the former and that they rear children as evidence of the latter. However, her second reason is a very deliberate attempt "to keep hold of the bodily aspect of existence—perhaps to grasp it overfirmly to keep it from floating away!" (p. 289).

Her approach carries with it the difficulty that although she uses Marx's method against his own naturalistic thinking about the division of labour within the family, the concept of sex pulls her arguments too strongly back to nature. However, this is a small point, easily met by the use of the cumbersome if precise notion of the sex-gender division of labour. What is more exciting is Hartsock's general approach to a distinctive feminist epistemology arising from women's experience/activities/labour within the world. While a main focus of her paper is the U.S. literature on motherhood and particularly the work of Nancy Chodorow and Jane Flax, she also discusses rather more briefly but illuminatingly women's unique labour and knowledge. Here she compares the knowledge available to (men) capitalists, (men) proletariats, and to all women regardless of class background:

> The focus on women's subsistence activities rather than men's leads to a model in which the capitalist (male) leads a life structured completely by commodity exchange and not at all by production, and at the furthest distinction from contact with material life. The male worker marks a way-station on the path to the other extreme of the constant contact with material necessity in women's contribution to subsistence. There are of course important differences along the lines of race and class. (Ruddick, 1980, p. 292)

She tellingly cites Marilyn French's account in *The Women's Room* of Mira reflecting on the significance of cleaning the loo:

> Washing the toilet used by three males, and the floor and the walls round it is, Mira thought, coming face to face with necessity. And this is why women are saner than men, did not come up with mad absurd schemes men developed; they were in touch with necessity, they had to wash the toilet bowl and floor.

My own arguments, begun in "Hand, Brain and Heart" (Rose, 1983), of how a feminist epistemology might be developed for the natural sciences, follow a common thread. My similarly menial story comes from listening to a radical male sociologist develop a social constructionist account of old age. I was sitting next to one of my feminist students, who worked with very elderly and acutely disabled people and with other women who also cared for them. She was beside herself with anger, "You can tell that he has never cleaned up a doubly incontinent old person." Just so; such sensuous activity would have constrained such insane social constructionism.

For reasons of space and time I will not recapitulate Hartsock's review of the object-relations school of feminist psychoanalysis, although I would note in passing that the U.S. work is much more securely grounded in a biosocial perspective, whereas its British sisterwork, perhaps because of its literary origins, is much less well integrated. From the differential experience of being mothered the male infant has to become the boy and man, to learn to abandon the material reality of that early experience. "Masculinity," she writes, "must be attained by opposition to daily life, by escaping the female world of the household to the masculine world of public life."

At this point in Hartsock's analysis, she speaks of a distinctive male experience that "replicates itself in both the hierarchical and dualistic institutions of a class society and in the frameworks of thought generated by this experience." Here, I think she merges the attack on Western culture and its oppressive dichotomies with her analysis of the male experience, where I would want to hold on, particularly in the case of the sciences, to the distinction between manual and mental labour that divides men as a group. Thus, while it may be true that masculinity is closer to the culture of death, it is only a particular stratum of men who advance that culture in evermore fearful technological form.

Sohn-Rethel (1978), for example, working in the tradition of the Frankfurt School, sought specifically to explain the social origins of the highly abstract and alienated character of scientific knowledge. Drawing on historical material, he suggests that abstract thought arises with the circulation of money, but goes on to argue that the specific alienated and abstract character of scientific knowledge arises out of the profound division of intellectual and manual labour integral to the capitalist social formation. Scientific knowledge and its production system are of a piece with the abstract and alienated labour of the capitalist mode of production itself. The Cultural Revolution in China, with its project of transcending the division of mental and manual labour, was seen by Sohn-Rethel, and indeed many or most of the new left, as offering a vision of immense historical significance. They saw within this social movement the possibility not only of transcending hierarchical and antagonistic social relations, but also as a means of creating a new science and technology, which was neither about the domination of nature nor of humanity as part of nature. Especially today, when the experience of the Cultural Revolution is problematic to assess, it is important to affirm our need of the project it represented.

Workers' movements in advanced capitalist societies shared in this affirmation of the knowledges derived from manual work (Cooley, 1979). Some projects, such as the Lucas aerospace workers alternative plan, represented the unity of the product of "hand" and "brain" as a new political and cultural goal for the workers' movement. Yet, even while these 1970s projects of the male workers' movement sought to produce new social products, their thinking often took for granted the caring labour necessary to bring the proposed new technology into practical use.

THE LABOUR OF LOVE AND ITS KNOWLEDGE

Feminism, aided by the crisis in the welfare state, not only named, and thus brought into visibility, the distinctive labour of women, but also insisted that we understand its doublesidedness, both as labour and as love. This combination of menial labour, often involving long hours, boring repetitive housework, and very complex "people work" with children, husbands, and dependent elderly people, has not been easy to unravel. It was right to insist that women's work was denaturalised, that it was integral to the division of labour and that much of it was socially and financially undervalued.

Skills acquired by women through practice within the home are systematically denied their social origins, whether they are utilised in paid or unpaid labour. In the new world factories of electronics, women's skill at microcircuitry and patience with repetitive tasks are seen unidimensionally as biological attributes; e.g., Elson and Pearson (1981); Sivanandan (1983). Within public caring labour, the relational skills of women, even where acknowledged as when a patient is said to need TLC (tender, loving care), still do not rate acknowledgement in terms of status or financial reward. In the domestic context, women are simultaneously praised for their "natural" nurturative qualities yet these qualities are seen as prescientific practices, awaiting the emancipatory certainty of scientific knowledge.

Experiential knowledge is thus dismissed and trivialised. At the same moment, one arrogant objectivising science seeks to instruct women in caring practices while another objectivising science sees them as inherently female. Not for nothing does the feminist cartoon say, "Well, if I get my instincts biologically I'm not having you tell me what to do!" The increasing scientisation of caring has undeniably eroded women's confidence, delegitimatising the knowledge they have gained individually and intergenerationally from the practice of caring. This vanguardist conception of science, a Bolshevism of the laboratory, appoints itself to lead women into emancipation, for when this objectivising science has mapped mothering definitively, men will be able to carry it out as well as women. It turns, however, as vanguardist theories do, on the denigration and disempowering of those it purports to aid. It refuses to understand how, if these are learned practices, they are transmitted. For Carol Gilligan (1982) by contrast, caring is in her terms, women's "intuitive ability [that] comes only with a certain sort of training" (p. 55).

Women themselves emphasise the accumulation of skill, the second baby is easier than the first, even though each infant is unique and requires a special and highly flexible response. The problem for women has been, rather, how to share and develop collective knowledge when experiential knowledge is dismissed as purely subjective. Studies of mothers consistently report that women's sense of self-confidence has been eroded by the scientisation of reproduction. Yet those areas where the organisation of caring is almost entirely in men's hands and claims to be guided by the achievements of science are precisely the areas where fad and fancy seem to have been freest, rather than in the prescientific practices of domestic caring. The sorry history of the biomedical sciences, above all psychiatry and gynaecology, provide all too many examples of women as victims of scientific fad and fancy. There are good arguments that the scientisation of domestic work has actually harmed practice. Waerness (1983) examines cookbooks and shows how those inspired by the latest scientific nutritional thinking led to unsound advice, while the practical guides offered by women cookbook writers still had a good deal going for them. She makes a parallel argument for cooking to that made by Donnison (1977), Ehrenreich and English (1973), and Versulen, Barber, and Allen (1976), for the replacement of the midwife by the, in reality more ignorant, man doctor.

ALIENATED AND NONALIENATED CARING

It is important in all forms of labour to insist that the experience of the unalienated form is located, however fleetingly, within the alienated, otherwise we have no means of conceptualising the future social relations and labour processes of a society that has overcome alienation. We capture it in our struggles, where in the social forms created through social movements we experience the anticipatory forms, the prefigurative forms of a new society. Novelists and poets, in their imaginative constructions of distopias and utopias, warn and inspire us with different, other societies. In analysing caring labour our double vision of such work, as both positive and negative, offers us a means of understanding why the work can offer tremendous satisfaction on one occasion and be the site of tremendously hostile and painful feelings on another, in which the person cared for confronts the caregiver as a hostile object. It also speaks of the pleasure and satisfaction found in the reciprocal care of feminism.

It was a problem both theoretically and empirically that, even where we tried to separate housework from peoplework, they continually merged. Caring, despite the best efforts of social work and psychotherapy, requires much more than the abstraction of words. We could feel in our heads, our hands, and our feelings the satisfaction of caring for someone, making someone content, finding all the little pieces of comfort that were important to that small child, that very elderly person: a mixture of words and silences, of favourite foods and drink, of hard work in cleaning up a wet or dirty bed, of special ways of doing things. Often tiring, it was satisfying, knowing that you had worked it round, you had taken care of them. All senses were involved; the person looked good, felt good, sounded good, smelled sweet. Yet, the pleasure at best did not just belong to the caregiver; it belonged to the cared for as well.

Alienated or unalienated, freely exchanged in reciprocal caring, given as a labour of love, enforced by an individual man or by the state, internalised by duty or the fear of gossip, women's caring labour is, and is much more than, the formation of feminine identity (Graham, 1983). As a profoundly sensuous activity, women's labour constitutes a material reality that structures a distinctive understanding of the social and natural worlds. Feminism has developed a strong sense of its potentiality in offering a transformative knowledge of the social world, yet, if it is to go beyond being a mere speciality within the social sciences, it has to claim the full strength of a feminist materialism that can overcome the old, oppressive dichotomy between the natural and the social.

BEYOND MASCULINE REALITIES—THE CLAIMS OF A FEMINIST THEORY OF KNOWLEDGE

I began this paper by asking, Is a feminist science possible? I described the achievements of feminist scholarship in its analysis of the present state of science. Feminism has rendered visible the exclusion of women from science; it has rescued from the oblivion of male history many of the women who have entered science; it has fostered the coming together of feminist scientists to rebut the claims of biological determinism; and it has recognised the theory and practice of the domination of nature as the specific feature of masculinist, bourgeois science. By discussing the nature of women's labour I tried to show how the exclusion of women and their knowledge from the practice and cognitive domain of science has come about.

A feminist epistemology derived from women's labour in the world must represent a more complete materialism, a truer knowledge. It transcends dichotomies, insists on the scientific validity of the subjective, on the need to unite cognitive and affective domains; it emphasises holism, harmony, and complexity rather than reductionism, domination, and linearity. In this, it builds on traditions that have always been present in science, though submerged within the dominant culture, and it joins hands with the critique of science as now practiced, which has developed within the new social movements.

Of course, as Elizabeth Fee (1983) argues elsewhere, a fully feminist science is not achievable outside a society that has not been fully transformed by the feminist project. But, just as the movement has been called into being by history, so it begins to develop its own here-and-now prefigurative forms. For history has called into being not only an unprecedentedly powerful feminism but also its sister, the ecology movement, whose agenda in many ways closely parallels that of feminism and whose struggles, together with those of feminism, touch the lives of men as well as women. Thus, while at a certain formal level of argument Fee is correct, the position does not enable us to explain such cultural products of feminism such as *Our Bodies, Ourselves* (1976). This book symbolises the flood of writing and talking about women's bodies and, like that flood, has been created out of shared experience. Here, it is important to stress that this celebration of shared experience added unquestionably new and qualitatively different biosocial knowledge. While women who have shared in this process recognise and value the new knowledge most highly, numbers of men—initially, particularly men health workers, who wanted to offer more sensitive care to women patients—also welcomed the new knowledge.

Nor is this listening and welcoming confined to the highly educated. Margaret, a miner's wife, describes one of the striking miners in Britain (North Yorkshire Women Against Pit Closures, 1985) who drives them to meetings

> We're always on about what we're going to do when the strike's over. How nothing'll ever be the same again. He listens patiently. "When this lot's over," he'll say, "I'm going into gynaecology or midwifery or something like that. I've travelled with you lot so often that I've got all the theory. When the strike's over, as I see it, I might as well get down to the practicals." (p. 64)

Periods of intense social struggle are, of course, also periods of intense creativity as well as suffering. Participants in struggle, of necessity, have to take one another's knowledge seriously and learn, and feel themselves learning, with unprecedented speed.

Nonetheless, in addition to celebrating and validating the new knowledges born out of women's collective resistance and struggle to gain power over their lives and bodies, there is simultaneously an attempt by feminists to propose, as I have done here, a distinctly feminist science and technology. As Sandra Harding (in press) pointed out in an illuminating paper, such a project can be read in two ways: as part of post-modernism, in which a feminist epistemology takes its place amongst a plurality of discourses, or as a successor science project, which claims to provide a better, truer picture of reality. Harding demonstrates that the contradiction between these two projects for a feminist epistemology is more apparent than real. Those who propose a distinctively feminist science share much of that successor science tradition about which Fee writes in this book (Chapter 3) and Rose and Rose (1971) wrote earlier; this claims to make over science, to locate it in proletarian values, Chinese, radical, and now, feminist values. Yet, there is a difference between those alternative epistemologies that are direct descendants of the Enlightenment tradition, those that want to claim both reason and objectivity as the historic allies of the oppressed masses, and those that still claim reason, but simultaneously give new meaning to the category of reason itself.

Like Harding and others, I want to sustain the feminist successor science project, not least because I share her political understanding that it "empowers all women in a world where socially legitimated knowledge and the political power associated with it are firmly located in white, Western, bourgeois, compulsorily heterosexual men's hands" (Harding, in press). Theoretically, this position enables us to understand the interconnections between those who argue for a plurality of discourses and for whom the main enemy is hegemony itself (Keller, 1985, p. 179) and the feminist successor science project. Not for the first time, feminist theory enables us to transcend dichotomous reasoning. Nonetheless, both issues of politics and theory make insistent claims on feminist thinking about science and technology, for we do so at a time when the need to transform these, and the society of which they are part, has never been more acute. In *Towards 2000*, Raymond Williams (1985) speaks of both the despair and the pessimism that afflict us when we think about the future, seeing the millenium itself as apocalypse, as nuclear holocaust, and also of the surprising energy and resilience displayed within contemporary socialism and of the confidence of those most committed to its values. For feminists, this sense both of dark times and well founded optimism is doubly true. Part of our energy comes from the empowering visions offered us by feminist science fiction writing—that fifth strand—that enables us to feel like Piercy's (1976) Dawn, not only do we "want to do something important" and move beyond masculinist conceptions of reality, but we can. Through her we feel that

> Someday the gross repair will be done. The oceans will be balanced, the rivers flow clean, the wetlands and forests flourish. There'll be no more enemies. No Them and Us. We can quarrel joyously with each other about important matters of idea. (Piercy, 1976, p. 328)

REFERENCES

Arditti, R. (1980). Women drink water: Men drink wine. In R. Arditti, P. Brennan, & S. Cavrak (Eds.), *Science and liberation*. Boston: Southend Press.

Balbo, L. (1983). Crazy quilts: Women's perspectives on the welfare state crisis. Mimeo, Milan. Italy: University of Milan.

Bkier, R. (1984). *Science and gender: A critique of biology and its theories on women*. Elmsford, NY: Pergamon Press.

Boston Women's Collective. (1976). *Our bodies, ourselves*. New York: Simon and Schuster (also 2d ed., 1985).

Brighton Women and Science Group. (1980). *Alice through the microscope*. London: Virago.

Carson, R. (1962). *Silent spring*. Boston: Houghton Mifflin.

Cooley, M. (1979). *Architect or bee*. Slough, England: Hand and Brain Publications.

Couture-Cherki, M. (1976). Women in physics. In H. Rose & S. Rose (Eds.), *Ideology of/in the sciences*. Cambridge, MA: Schenkman.
Daly, M. (1979). *Gyn/ecology: The metaethics of radical feminism*. London: The Women's Press.
Donnison, J. (1977). *Midwives and medical men*. London: Heinemann.
Ehrenreich, B. & English D. (1973). *Midwives, witches and nurses*. New York: The Feminist Press.
Elson, D. & Pearson, R. (1981, Autumn). Nimble fingers make cheap workers: An analysis of women's employment in third world export manufacturing. *Feminist Review, 9*.
Fee, E. (1983). Women's nature and scientific objectivity. In M. Lowe & R. Hubbard (Eds.), *Woman's nature: Rationalisations of inequality*. (Athene Series). Elmsford, NY: Pergamon Press.
Firestone, S. (1971). *The dialectic of sex*. London: Cape.
Gilligan, C. (1982). *In a different voice: Psychological theory and women's development*. Cambridge, MA: Harvard University Press.
Graham, H. (1983). The labour of love. In J. Finch & D. Groves (Eds.), *The labour of love*. London: Routledge and Kegan Paul.
Harding, S. (in press). *The instability of the analytical categories of feminist theory. Signs*.
Hartsock, N. (1983). The feminist standpoint. In S. Harding & L. Hintikka (Eds.), *Discovering reality*. Dordrecht, The Netherlands: Reidel.
Hubbard, R., Henifin, S., & Fried, B. (Eds.). (1980). *Women look at biology looking at women*. Cambridge, MA: Schenkman.
Hubbard, R., & Lowe, M. (Eds.). (1979). *Genes and gender*. New York: Gordian Press.
Keller, E. F. (1977). The anomaly of a woman in physics. In S. Ruddick & P. Daniels (Eds.), *Working it out*. New York: Pantheon.
Keller, E. F. (1983). *A feeling for the organism: The life and work of Barbara McClintock*. New York: W. H. Freeman.
Keller, E. F. (1985). *Reflections on gender and science*. New Haven, CT: Yale University Press.
Kelly, A. (Ed.). (1981). *The missing half*. Manchester, England: Manchester University Press.
Land, H., & Rose, H. (1985). Compulsory altruism for some or an altruistic society for all. In P. Bean, J. Ferris, & D. Whynes (Eds.), *Defence of welfare*. London: Travistock.
Leibowitz, L. (1978). *Females, males and families: a biosocial approach*. Cambridge, MA: Duxbury Press.
Leiss, W. (1972). *The domination of nature*. Boston: Beacon Press.
Lowe, M., & Hubbard, R. (Eds.). (1983). *Woman's nature*. (Athene Series). Oxford, England: Pergamon Press.
Mass, B. (1976). *Population target: The political economy of population in Latin America*. Brampton, Ontario: Charters Publishing Co.
Merchant, C. (1980). *The death of nature*. New York: Harper and Row.
Narek, D. (1970). A woman scientist speaks. In L. Tanner (Ed.), *Voices from women's liberation*. New York: Signet.
North Yorkshire Women Against Pit Closures. (1985). *Strike: 84–85. People's History of Yorkshire*. Leeds, England: Author.
Osen, L. (1974). *Women in mathematics*. Cambridge, MA: MIT Press.

Piercy, M. (1976). *Woman on the edge of time*. New York: Knopf.
Pinchbeck, I. (1930). *Women workers and the industrial revolution 1750–1850*. London: Virago (republished 1981).
Polanyi, M. (1967). *The tacit dimension*. Garden City, NY: Anchor Books, Doubleday.
Rich, A. (1974). *Of woman born*. London: Virago.
Richter, D. (Ed.). (1983). *The path to liberation*. New York: UNESCO.
Rose, H., & Rose, S. The radicalisation of science. In R. Miliband & J. Saville (Eds.), *The socialist register*. London: Merlin. (Reprinted in Rose, H., & Rose, S., (Eds.), 1979, *Ideology of/in the Natural sciences*.)
Rose, H., & Rose, S. (Eds.). (1979). *Ideology of/in the natural sciences*. Cambridge, MA: Schenkman.
Rose, H. (1983). Hand, brain and heart: Towards a feminist epistemology for the sciences. *Signs, 9*, 73–90.
Rossi, A. (1965). Why so few? *Science, 148*, 1196.
Ruddick, S. (1980). Maternal thinking. *Feminist Studies, 6*, 342–367.
Sayers, J. (1982). *Biological politics*. London: Routledge and Kegan Paul.
Sayre, A. (1975). *Rosalind Franklin and DNA: A vivid view of what it is like to be a gifted woman in an especially male profession*. New York: W. W. Norton.
Schmidt, A. (1971). *The concept of nature in Marx*. London: New Left Books.
Science for the People Collective. (1977). *Sociobiology as a social weapon*. Ann Arbor, MI: Burgess.
Sivanandan, A. N. (1983). *A different hunger*. London: Pluto.
Sohn-Rethel, A. (1978). *Intellectual and manual labour: A critique of epistemology*. London: Macmillan.
Tobach, E., and Rossoff, B. (Eds.). (1978). *Genes and gender*. New York: Gordian Press.
Versulen, U., Barber, D., & Allen, S. (Eds.). (1976). Midwives, medical men, and "poor women laboring of child": Lying-in hospitals in 18th-century London. In H. Roberts (Ed.), *Women, health and reproduction*. London: Routledge and Kegan Paul.
Waerness, K. (1984). The rationality of caring. In H. Holter (Ed.). *Patriarchy in a welfare society*. Bergen, Norway: University of Bergen Press.
Walby, S. (in press). *Patriarchy at work*. Cambridge, England: Polity Press.
Walkerdine, V. (1981). *Girls and maths: The early years*. Bedford Way Papers 8. London: University of London.
Weisstein, N. (1977). Adventures of a woman in science. In S. Ruddick & P. Daniels (Eds.), *Working it out*. New York: Pantheon.
Williams, R. *Towards 2000*. Hammondsworth, England: Penguin.
Women and science and society [special issue]. (1978). *Signs, 4*(1).
Woolf, V. (1938). *Three guineas*. London: Hogarth Press. Reprinted, Hammondsworth, England: Penguin (1977).
Ziman, J. (1969). *Public knowledge*. Cambridge, England: Cambridge University Press.

Chapter 5

Primatology Is Politics by Other Means*

Donna Haraway

> Rouse ye, my people, shake off torpor, impeach the dread boss monkey and reconstruct the Happy Family.**

INTRODUCTION: SCIENCE AND STORYTELLING

The science that is the study of monkeys and apes, primatology, is a major area of feminist concern about the tangled relations of gender, knowledge, and power. This is true both because women have contributed significantly to these areas of biology and anthropology and because these sciences are important in debates about human, perhaps especially female human, nature. Men and women are primates; we consider ourselves to be animals in the taxonomic order *primate*, at least since the 1758 edition of Carl Linnaeus's *Systema Naturae*. From the perspectives of the natural and social sciences constructed within this framework since the 18th century, the other primates have a series of special relations to human beings. They are privileged beings for understanding "nature" and "culture," among the principal analytical categories Western people have used to theorize their histories and experience. Monkeys and apes are mirrors for human beings in our aspect as animals. They are constructed to tell us what is "beneath," "at the heart of," or "outside" of language-using animals; that is, ourselves.

So the nonhuman primates are seen to exist at a crucial boundary between

*This chapter has been edited from Donna Haraway's article, "Primatology Is Politics by Other Means," *Philosophy of Science Association 1984* (Vol. 2). East Lansing, MI: Philosophy of Science Association. Copyright 1984 by The Philosophy of Science Association. Used with permission.

**Mark Twain (March 5, 1867). Barnum's First Speech in Congress. *New York Evening Express*, as quoted in Harris (1973, 190–191).

animal and human. And, although that boundary has changed repeatedly throughout history, it is drawn so as to appear to be a natural line. While primatology is an important discourse and social practice for outlining (literally, sketching the boundaries of) a putatively universal human nature, including the nature of "woman," not all women (or men) have contributed to these domains of knowledge. The "scientific" way of looking at monkeys and apes is thoroughly, historically, and culturally specific; it has been inconceivable to most men and women on the planet to develop this kind of knowledge.

Until very recently in the history of primatology, virtually no women had the status of scientist in this field. Subsequently, only European and Euro-American women, with very few exceptions, have watched monkeys and apes professionally. These women have made a major difference in scientific constructions in primatology of what it means to be a female animal, and so of what it means to be a man or woman in societies for which the social construction of animal is part of the social construction of human. In other national scientific traditions in primatology, for example in India and Japan, there have been notably few women scientists. Though it should not be underestimated, racism in science is only one reason for this racial and national character of primate scientists. It has made sense to white industrial people, and perhaps especially U.S. white women, to study these animals in ways that have not made sense to other people.

How and where do people *see* nonhuman primates? In Europe and North America, simians can be seen alive in zoos, stuffed in natural history displays, illustrated in school books, photographed for popular magazines, or projected in movies—all settings accessible to people without the status of scientist. Monkeys and apes in what are called *laboratory colonies* can be watched by scientific staff, lab janitors, animal technicians, and an occasional journalist or filmmaker.

But monkeys and apes do not generally live in Europe and North America. They live overwhelmingly in the tropics, in Africa, Asia, and Latin America; that is, they live mostly in the Third World. Those animals with an almost magical status for late-industrial capitalist people live in a kind of distant dream space produced out of the history of colonialism, symbolized by the mountain gorilla or the chimpanzee in the heart of Africa. This dream space is perhaps better conveyed by the television nature special, like those produced by the National Geographic Society (sponsored for over 10 years by Gulf Oil Company) than by the scientific research paper. But the two genres of primatology are intricately linked. To watch wild monkeys and apes, to find that particular "nature" that exists outside "culture"—and, therefore, has special implications for modern Western theories of human nature and society—is to enter into the history of Western expansion and colonialism. Indeed, the special symbolic status of "wild" animals is part of the history of colonial discourse. The history of wild animals is intimately part of the history of race, sex, and class in a world capitalist system.

So Westerners have access to monkeys and apes only under specific symbolic and social circumstances. Although systematically obscured and denied in the ideologies of culture-free, objective science, these circumstances matter to the fundamental nature of the sciences produced. Europeans and North Americans must generally travel long distances and undergo the special experience called *field work* in order to write accounts of these animals' lives and their meanings for other men and women. The experience of "nature," able to be perceived as free of human agency, is expensive and hinges on the exercise of institutional power. In addition, perhaps because they have been forced by an unusual permeability of their sciences to "outside" political and popular contestation and debate, field primatologists are peculiarly aware of and troubled by the patent differences in the primatologies authored by men or women, Japanese or Dutch nationals, British ethologists, or North American physical anthropologists. It is very hard to stabilize the truth about monkeys and apes.

It is the social authority to write scientific accounts of what are called *wild primates* that concerns us in this essay. A scientist is one who is authorized to name what can count as nature for industrial peoples. A scientist "names" nature in written, public documents, which are endowed with the special, institutionally enforced quality of being perceived as objective and applicable beyond the cultures of the people who wrote those documents. How have white feminist women contested for meanings of nature in these social and symbolic conditions?

In the historical, philosophical, and social studies of science, it has become commonplace to note that "facts" depend on the interpretive framework of theory, and that theories are loaded with the explicit and implicit values of the theorizers and their cultures. Thus, all facts are laced with values. Clearly, I have set up my story of primatology to explore the connections of facts and values in an area of biology and anthropology that functions simultaneously as natural science, political theory, and science fiction.

But *values* seems an anemic word to convey the multiple strands of meaning woven into the bodies of monkeys and apes. So, I prefer to say that the life and social sciences in general, and primatology in particular, are story-laden; these sciences are composed through complex, historically specific storytelling practices. Facts are theory-laden; theories are value-laden; values are story-laden. Therefore, facts are meaningful within stories.

The story quality of the sciences of monkeys and apes is not some pollutant that can be leached out by better method, say by finer quantitative measures and more careful standards of field experiment. Improved method matters, as it does in any human craft. Improved method, including improved ways of asking questions, is a collective achievement, and stories are not equivalently good. The point is not that one account of monkeys and apes is as good as another, since they are all "merely" culturally determined narratives. Rather, I am arguing that the struggle to construct good stories is a major part of the craft. There would be no primatology without skillful, collectively contested stories. And, there would be no stories, no questions, without the complex webs of power, including the tortured realities of race, sex, and class—and including people's struggles to tell each other how we might live with each other.

The sciences have always had a utopian character. In their efforts to *describe* the world, to understand how it actually "works," scientists simultaneously search out the limits of possible worlds. What determines a "good" story in the natural and social sciences is partly decided by available social visions of these possible worlds. Description is determined by vision; facts and theories are perceived within stories; the worlds for which human beings contest are made of meanings. Meanings are tremendously material forces—much like food and sex. And, like food and sex, meanings are social constructions that determine the quality of people's lives.

That is why the intersection of feminist and colonial discourse in the sciences of monkeys and apes matters. Monkeys and apes have been enlisted in Western scientific story telling to determine what is meant by human: what it means to be female, to be animal, to be other than man. As white women achieved the social and symbolic power conferred by scientific degrees and the ability to watch nonhuman primates as principal investigators, rather than zoo visitors, they brought with them histories, experiences, and world views that reconstructed basic stories in primatology. They changed the facts of nature by changing the visions of possible worlds, and it has been hard, complicated work. Primatology is the scene of a feminist scientific revolution, one that has changed the way both men and women practice their science, at least sometimes and in areas of considerable importance.

The reconstituted nature, however, remains deeply Western: deeply marked by the logics of nature and culture, by Western searches for the self in the mirror of a subordinated other, by the constantly repeating origin stories that ground Western political culture. I learned that primatology is simian orientalism at the same time that I learned how deeply it mattered that some women contested successfully for good scientific stories in primatology. This essay is a tentative effort to articulate a feminist history of primatology in view of that joint consciousness.

Feminist contests for authoritative accounts of evolution and behavioral biology are not simply alternatives, but equally as biased as the masculinist stories so prominent in the early decades of the field. To count as better stories, they have to better account for what it means to be *human* and *animal.* They have to offer a fuller, more coherent vision, one that allows the monkeys and apes to be seen more accurately. That is one of the rules of the scientific game, one of the great strengths and troubling universalisms built into the claims of Western science (Harding, 1985). But what will count as more accurate, fuller, more coherent?

Rarely will feminist contests for scientific meaning work by replacing one paradigm with another, by proposing and successfully establishing fully alternative accounts and theories. Rather, as a form of narrative practice or storytelling, feminist practice in primatology has worked more by altering a "field" of stories or possible explanatory accounts, by raising the cost of defending some accounts, by destabilizing the plausibility of some strategies of explanation. Every story in a "field" alters the status of all the others. The total interrelated array of stories is what I call a *narrative field.* The following essay deliberately plays on the many meanings of the term *field*: as the place where people go to see wild monkeys and apes, to do "field studies"; as the set of scientific discourses constituting primatology, especially biology, psychology, and bioanthropology; as the dynamic web of stories and possible meanings; as the many complex spaces where meanings are contested and stabilized for a time through the productive relations of knowledge and power.

Altering the structure of a field is quite different from replacing false versions with true ones. To construct a different set of boundaries and possibilities for what can count as knowledge for everyone within specific historical circumstances is a radical project. Feminist science is not biased science, nor is it disinterested in accurate description and powerful theory. My thesis is that feminist science is about changing possibilities, not about having a special route to the truth about what it means to be human—or animal. In that sense, primatology is a genre in feminist theory.

The essay itself works as a story, beginning with "The Field: Origins," where organizing axes of Western stories are examined. I am especially interested in the ways the feminist analytical distinction between sex and gender is a version of the Western dynamic dualism, nature–culture. The issue matters because much of women's energy in primatology has been directed to restructuring what it can mean to be "female." Women have contested for the meaning of the gender "woman" by reconstituting what female sex means from the point of view of life sciences. The next section, "The Jungle: Spaces," sketches briefly a history of women who practice primate studies. The final section, "The Text: Representations," explores a particular contest in paleoanthropology and sociobiology for the meaning of sex in human, especially human female, evolution. The essay works by a layering of meanings and deliberate punning; the attempt is to provide a convincing account of the history of science that is simultaneously political theory, science fiction, and sound scholarship. This essay is itself a kind of primatology, only the primates observed and worked into narratives they did not invent are those who watch monkeys and apes for a living.

THE FIELD: ORIGINS

Adam and Eve, Robinson Crusoe and Man Friday, Tarzan and Jane: these are the figures who tell white Western people about the origins and foundations of social life. The stories make claims about "human" nature, "human" society. Western stories take the high ground from which man—impregnable, potent, and endowed with a keen vision of the whole—can survey the field. The sightings generate the simultaneously aesthetic and political dialectic of contemplation and exploitation. These qualities, mystic love of nature and the desire for limitless instrumental power, are the distorting mirror twins, deeply embedded in the history of science.

But the moment of origin in these Western stories is solitary. Adam was alone, Robinson was alone, Tarzan was alone; they lacked human company. Each couple promised to be the solution to the illogical insufficiency of a rational autonomous self. But each couple proved to be fraught with the contradictions of domination that provided the narrative materials of the West's accounts of its devastating collective history. The tragedy of the West is rooted in number: one is too few and two are too many. To be one should mean to be unified, whole; that should be enough, yet it is lonely. But all human community involves difference. Difference is a challenge to autonomy, to wholeness. Memory, always about the origin, is about a lost oneness imagined as sameness. The telos or goal is about perfect union. The origin and the purpose, then, both are about the desire to be One. The process of mediating the beginning and the end, called *history*, is a tale of escalating domination toward the apocalypse of the final transcendence of difference. Until the end, difference is dialectical and dynamic antagonism; at the end, difference is transubstantiation and communion. This is the essential structure of masculinist eros so deeply embedded in Western stories, including the natural sciences.

This essay in the history of zoology and bioanthropology is part of a worldwide oppositional effort, rooted in social movements like feminism and antiracism, to retell these stories as a strategy to break their power. My thesis is that the scientific practices and discourses of modern primatology participate in the preeminent political act in Western history: the construction of Man. This construction evokes collective deconstruction. Primatology is politics by other means, and women's place is in the jungle, arguing the nature of beginnings and ends. The life stories of monkeys and apes are industrial and postindustrial versions of the past and the future of them and us. Primatology is a complex scientific construction of self and other, culture and nature, gender and sex, human and animal, purpose and resource, actor and acted upon. This scientific field constrains who can count as "we." Mind and sex provide most of the drama. Primatology is also compelling soap opera.

Adam ruled Eve in the foundation of compulsory heterosexual reproductive politics. He ruled her in retribution for her disruption of the boundaries that made the Garden possible. Just as Milton's *Paradise Lost* marked the retelling of Adam and Eve's story at the "moment of origin" of the scientific revolution, Protestant Christianity, capitalism, and Western expansion, it is possible to see current evolutionary scenarios as retellings of the first family, the first "we," at the "moment of origin" of multinational capitalism, secular humanism, the information sciences revolution, and the emergence of the Third World. Many of these scientific versions of sexual politics, of origins of "the family," normalize compulsory heterosexual reproductive politics with a verve that might have made even Jehovah pause.

One way of looking at storytelling by women primatologists is to see them as "Milton's daughters," whose materials are necessarily the inherited stories that mark the biological category "female," as well as actual women, as *other* (Gubar & Gilbert, 1979; Haraway, 1981; 1983b). Female and women, the marked categories, are inflected, linguistically and socially. To be the unmarked category is to be the norm; to be the marked category is to be the dependent variable. What is a family without dependent, marked, normalized members? That should not be a rhetorical question.

This essay explores what is at stake when "female" is both the object of study and the condition of the observers in contemporary contests for authoritative origin stories. Clearly, origins here are not about specific historical, or even prehistorical, events and durations. The time of origins is mythical, and the tension between mythical and other kinds of time is part of the structure of Western scientific discourse (Fabian, 1983).

Robinson Crusoe subordinated his companion in the drive to rationalize time and space on the island; male union in love and equality was a tantalizing dream, but only if the boundaries of mind and body essential to order could be sustained. Nancy Hartsock (1983a, 1983b) called this order *abstract masculinity*. Michel Tournier (1972) redid Defoe to erase the crushing rationality of Robinson's boundaries. The sociologist and philosopher of science, Bruno Latour (1984), used Tournier's version of Crusoe to contest the boundaries separating "science" as a sacred center protected from the polluting arenas of politics. Latour gave me the title for this paper, as he appropriated von Clausewitz's famous, early 19th century counsel ("war is politics by other means") to craft a slogan for the social studies of science: "La science, c'est la politique continuee par d'*autres* moyens" (Latour, 1984, p. 257). Latour stresses that this view does not "reduce" science to politics, to arbitrary power rather than rational knowledge; those are not the stakes, but a mystifying dichotomy. Precisely, his is an argument against reduction of any kind and for attention to just what the "other means" are. In primatology the means centrally include narrative strategies and the social power to deploy them for particular audiences: "Tout se negocie" (Latour, 1984, p. 183). It should not be surprising that one woman primatologist intent on destabilizing meanings of social dominance and reproductive politics in baboon society has collaborated with Latour in analyzing the structure of origin stories (Latour & Strum, 1983).

As for Tarzan, his humanity seemed to hinge on his renunciation of the Garden; his final lordship had to be over self, a regime mediated by the civilized Jane. But Tarzan and Jane happily failed where Adam and Eve succeeded in making history and generating the peoples of the Book, the tragic subjects of salvation history and monotheisms. At least on TV in the 1950s, the former pair returned to the jungle, never properly married and with children of dubious provenance, one of whom was a chimpanzee. There is a hint of possibility of sociality without domination in this intriguing tale. It hinges on how you think about parenting an ape. In the 1950s on American television, it was radical fare.

But there are many versions of Tarzan; and, in the latest, the popular movie *Greystoke*, Tarzan and Jane were separated again, suspending their possibility of progeny who could bridge the distinction between human and animal. Witness to the threatened dissection and subsequent murder of the old male ape who protected him in Africa, the legitimate Lord Greystoke cried out in front of the British Museum, "He was my father," claiming the ape as his patrimony and suspending his promising betrothal to Jane.

A curious touch is that the natural-seeming simian father is a *simulated* ape, one who was born in film techniques, make-up artistry, the pedagogy of teaching human sign language to apes, and scientific field studies of gorillas and chimpanzees that (who) teach ape vocal and gestural communication to humans. Jane, too, is of dubious parentage, the American ward of the senior Lord Greystoke. Late 20th century versions of nature are more about simulacra than about originals (Baudrillard, 1983). These are stories about copies superior to originals that never existed. Plato's forms have given way to cyborg information, perfection to optimization. All the versions are about the problem of connection of hostile, but perversely echoing/reflecting/twinned, poles within social relations and culture myths based on dualisms.

It is not unimportant to this essay that Jane is always the civilized pole in the Tarzan stories; the gender "woman" easily carries the meaning of culture to the primitive pole, the gender "man". Nature-culture and feminine-masculine lace into networks with each other; the terms do not relate as isomorphisms or unidirectional parallels. That is, the paired distinctions feminine–masculine, body–mind, nature–culture, animal–human, and so on are systematically related to each other, but in many ways. Feminist analysis has frequently erred in assuming that the equation of woman with nature, animal, dark, etc., is the only relationship built into the series of dual terms. These dualist axes are story operators, ways of structuring relationships. They are not static ascriptions. Geometries organized around the dualisms of nature–culture and sex–gender structure the narrative of human relation to the animals and much else in primatology. Redistributing the narrative field by telling another version of a crucial myth is a major process in crafting new meanings. One version never replaces another, but the whole field is rearranged in interrelation among all the versions in tension with each other. Destabilizing an origin story is perhaps more powerful in the deconstruction of the history of man than replacing it with a more progressive successor.

Restructuring a field held together by its tensions is how primatology works to produce meanings around sex and gender. My story does not rest on substitutions of true versions for false, feminist for masculinist, scientific for ideological accounts. Primatology as a storytelling practice works by another process, another mechanism in the political contest for meanings. Many kinds of activities can restructure a narrative field, including practices for recording data, publishing patterns, favored animal models, a women's movement, developments in adjoining sciences, complexities in conservation politics, and new nationalist governments in east Africa (Beer, 1983; Landau, 1984; Nash, 1982, pp. 342–378; Strathern, 1980, 1984).

In the founding Western stories, each autonomous self was a man; indeed, each autonomous self was Man. But politics is about a "we." Politics only exists where there is more than one voice, more than one reality. Politics is about difference: its recognition, negotiation, suppression, constitution, exaltation, impossibility, necessity, scandal, and legitimacy. Gender is also about difference; it is the politics of the socialization of sex. Gender is the politics of an ordered, collective (but hardly shared and not exclusively public) world built from the profusion of differences originally *constructed* as sex. It is important not to make the mistake of thinking sex is given, natural, biological, and only gender is constructed and so social. Biology is an analytical discourse, not the body itself; and biological sex is an object of knowledge and practice crucial to power in the modern West (Foucault, 1978).

And finally, the West is about difference; it is the politics of the civilization of the "primitive," the domination of nature by culture. Culture appropriates nature in Western founding stories, just as gender is the social appropriation of sex as resource for social action. This fundamental relationship is built into our notions of causality, human nature, history, economics, etc. (Strathern, 1980). The one appropriates the other; from Aristotle to Hegel to Sartre, there has been no disagreement about basics. Gender and the West are eminently political constitutions because they order the differences central to the possibility of a collectively recognized and enforced reality. Race and sex are the chief bodily products of these political labors. The rule of order in these hard myths is the dialectical rule of act and potency, mind and body. Primatology is a complex scientific practice for simultaneously discovering and constructing natural–technical objects of knowledge within an epistemic field structured by sex–gender and by the West and its others.

The unanswered question in these politics is whether difference can be ordered by something other than deadly opposition without falling into the covert dominations of functionalism and organicism, both of which have been attractive ideologies for feminist analysts. Functionalism and organicism do not remove the principle of domination; they only remove the drama of dialectical and apocalyptic opposition. That merely makes domination mundane and boring. Functionalism "normalizes" domination, literally. It provides an analytical method for ordering difference into wholes organized by the hierarchical division of labor, in which everything becomes a "part" with a role and a place. Difference is turned into variation, described by a range of graphic and statistical tools, which work as both social metaphors and actual tools for enforcing certain ranges of variation and cutting the tails off of others. That is why functionalism has been the ruling approach in modern life and human science; it is the form of analysis proper to the knowledge–power relations of liberal society.

Organicism, or the dream of natural (unforced) community, is an honored possibility in Western religious, philosophical, and scientific tradition. For many radical and antiliberal thinkers, including many feminists considering modern sciences and technologies, it has also appeared to be an alternative to both antagonistic opposition and to regulatory functionalism. It is easy to forget that organicism is a form of longing for a spontaneous and always healthy body, a perfect opposite to the technicist and reductionist boogey man, often imagined as a scientist making bombs. What else could scientists be doing? Organicism is the analytical longing for a natural body, for purity outside the disruptions of the "artificial." It is the reversed, mirror image of other forms of longing for transcendence. Organicism is perhaps more dangerous for feminists than obviously abstract forms of masculinist transcendence, which do not hide their danger to ordinary, finite communities by claims of being fully embodied and natural.

Many strands within feminist theory are attempts to articulate a politics of shared and partial realities that value serious difference. Feminism must be opposed to holist organicisms if it is to avoid logics and practices of organic domination. Organicism, in science or politics, will not destabilize the story of man. Simultaneously, feminism must affirm the hope and partial reality of community. Perhaps, modern Western feminists might fruitfully imagine such bodies and communities in metaphors, as hybrids of organism and machine, of animal and human. The sciences and science fictions of the late 20th century are full of these cyborg metaphors, waiting to be explored for feminist possibilities (Haraway, 1985). My hope is that cyborgs relate difference by partial connection rather than antagonistic opposition, functional regulation, or mystic fusion. Curiously, nonhuman primates, and other biological objects of knowledge, have been retheorized as cyborgs in the late 20th century.

But, at least as many identifications and unities must be broken down as rebuilt, and any new construction requires *both* belief and disruption. Curiously, the set of social practices and discourses in Western culture that seem most to require these dual relationships to action and knowledge are the natural sciences. Cynicism is fruitless, worse than the wrong pH or cruel cage design for producing conditions yielding understanding of animals. Cynicism is another name for ideologically rigid objectivity. Faith seems more promising, full of possible connections to a "real" world. But "connections" can breed identifications, appropriations, and illusions of wholeness. These matters are played out in detail in the field studies of monkeys and apes. Primatology as a field of contest provides some intriguing patterns of political thinking about identity, association, and change. Most positions within the U.S. white women's movement are replicated, or better negotiated, within primatological discourse. Primatology is a genre of political discourse about the question of community.

Like other major systems of myth and political theory in Western storytelling tradition, primatology starts from a unit, a one, and tries to generate a whole, a we. And as in these other stories, the narrative tension in primatology comes from the drama of the dialectics of domination, the scandal of difference. Primatology is a utopian project, close to the heart of Western political theory. The sciences that study monkeys and apes are inherently about origins, about the nature of things. Even the name of the order, given by Linnaeus in 1758 in the eighth edition of the *Systema naturae*, means "first." The primate order has never been stable, whether the debates were between Huxley and Owen on evolution in the 19th century or between Adrienne Zihlman and Owen Lovejoy on bipedalism and reproduction in the 20th (Lovejoy 1981, 1984; Zacharias 1984; Zihlman & Lowenstein 1983).

Primatology is also a period in salvation history, deeply immersed in barely secular versions of the Garden and the Fall. It is also a discourse on first principles. So as utopian project, story of origins, secular version of salvation history, and discourse on philosophical first principles, primatology is a contested field within Western cultures for defining what it means to be human. As such, it is a major social practice for Western 20th century people to construct and negotiate the boundaries between human and animal, gender and sex, West and other, culture and nature, whole and part.*

*I am not considering here the central strand of Japanese and Indian primatology. Similarities and differences should be read in the context of specific founding myths and late industrial social relations. A comprehensive comparative study has been undertaken by Pamela Asquith of Calgary University (Asquith, 1984). The emergence of Third World primatologies, intimately connected to the political economic role of "vermin" and "wildlife," is another unexamined topic.

These boundaries are not in phase with each other on the map of the primate body but interact in every imaginable way, perhaps most often synergistically. The boundaries are somewhat analogous to the chakras of other bodily maps; they sustain localized bodily interventions but do not reveal the secret of any "real" physical presence. The field of negotiation of these boundaries—sustained by, but not reducible to (even "in the last instance"), material social relations of late 20th century systems of race, class, and sex—cannot help but command the passions of those of us with stakes in the proper (i.e., specific historical) constitution of human nature. I do not know anyone who does not have some stake in this territory. To the dismay of professional primatologists, constantly trying to license the practice of discourse on monkeys and apes, the affairs of these animals are a popular matter. The professionals acknowledge this fact repeatedly when they write and in technical journals cite as key points of their arguments hybrid texts like *Chimpanzee Politics* and *The Woman That Never Evolved* (de Waal, 1982; Hrdy, 1981). Both books are simultaneously contributions to the scientific literature, political arguments, and gripping fiction.

Readers older than 15 will also recall a host of earlier entries into the fray, like *Naked Ape* and *The Imperial Animal* (Morris, 1967; Tiger and Fox, 1971). Many of these books are now written by women scientists, whereas none of them were before the post-World War II period in primatology. Jane Goodall's *In the Shadow of Man* (1971) is a landmark in ape gender politics. I doubt that Goodall intended the irony I hear in that title, but her book may be read fruitfully as a chapter in the "reproduction of primate mothering" (Chodorow, 1978). Primatologists love the fray; primatology is full of exuberant action and the desire to narrate. And the "actors" include animals and people, laboratories and books, and a great many other categories of resources for crafting political orders (Callon & Latour, 1981; Latour, 1978). If there is a political unconscious, there is surely one horizon of it generating primatology as a socially symbolic act. (Jameson, 1981, pp. 77–89).*

---

*Jameson takes seriously Levi-Strauss's suggestion that "all cultural artifacts are to be read as symbolic resolutions of real political and social contradictions" (Jameson, 1981, p. 80). Jameson also points out that this proposition requires serious "experimental verification." However, Jameson's notions of a symbolic unconscious and his description of three phases of analysis, especially his focus on the symbolic act and the ideologeme ("the smallest intelligible unit of the essentially antagonistic collective discourses of social classes," p. 76), are rich for a reading of primatology. It is essential to reconstruct the notion of social classes to problematize the collectivities "women" and "men."

Life and Human Science

Primatology is also a branch of modern biology and anthropology, and, as such, is subject to the structuring of the life and human sciences within the webs of knowledge and power that make the sciences possible. Although there is no contradiction between this characterization and those asserting the mythic and political nature of the sciences of monkeys and apes, there is tension. Like the boundaries between nature and culture, sex and gender, animal and human, the scientific and mythic characters of primate discourses are not quite in phase. They evoke each other, echo each other, annoy each other, but are not identical to each other. Science and myth neither exclude nor replace each other; they are versions of each other. In the 20th century in the United States, they structure each other. Reducing science to myth or vice versa would obscure precisely the field of gender politics—and much else—which I regard as real and interesting. Reduction is rarely a very rich explanatory strategy, least of all when the goal is to evoke mediating strands and complexities tying together social-technical-symbolic life across sacred boundaries. Reading primatology is itself an exercise in boundary transgression.

And, to make matters even more tense, the life and human sciences are in a state of war, as fits any set of mythic twins, virtual images of each other. Biology, a natural science whose practitioners tend not to see themselves as interpreters but as discoverers moving from description to causal explanation, and anthropology, whose practitioners tend to argue their authority is the fruit of interpretation, set up a difference that structures primate science. In primatology, the stakes of the conflict are mundane: publications, jobs, status hierarchies among monkey watchers, preferred metaphors, explanatory strategies, favorite graduate schools, versions of histories of the discipline, etc. Until recently, the stakes have also been gender of monkey watcher: many more women primatologists originally came from anthropology than from the biological disciplines, a matter of no little consequence in the sociobiology and behavioral ecology debates. The contest is for the allowable meanings of "adaptation." The anthropologists have inherited the storytelling strategy rooted in structural functionalism deeply tied to social and cultural anthropology, while the biologists have inherited the storytelling strategy of positivism and empiricism deeply tied to the hegemonic authority of physical sciences. Both inherit versions of political economy. For primate anthropology, this has meant focus on the division of labor, role theory, and notions of social efficiency. Biology has dwelt on market analysis, econometrics, investment strategies, and life insurance demographic techniques. One looks more for social role and functional integration, the other for game theory calculations and cost/benefit analysis simulations (e.g., Fedigan, 1982; Hrdy, 1981).

Biology

First let us take an unconscionably brief look at the structuring of knowledge in the life sciences. From the late-18th and early-19th centuries through the mid-20th century, nature has been reconstituted as a system of production/reproduction and communication. The life sciences have been crucial to this very material transformation of objects of knowledge and practice. Nature has become an expanding system in which the rational control of the product of expansion operates as mind to body. Malthus was no fool; neither was Adam Smith nor Charles Babbage nor Henri Milne-Edwards nor Charles Darwin. They all understood what the division of labor did to nature and used this not-so-covertly hierarchical principle for all it was worth. It turned out to underlie exchange value, to ground the common coin or currency of life, the natural economy. Functional explanation in biology of the body's differentiation into specialized subsystems is subordinated to explanatory strategies drawn from the market, to investment and cost–benefit explanations built deeply into the theory of natural selection. How far away scarcity assumptions and market constraints lie sets the boundaries for debate about levels of explanation in biology, including primatology. Fewer or more "degrees of freedom" are at stake. But, in biology, ultimate stakes are staying in the game, replication, differential reproduction.

From the point of view of life as an expanding system, two "subsystems" or functional specializations structure biology in a special way. Since the constitution of life as a "natural–technical object of knowledge," without caricaturing too much, it is possible to tell the history of biology in terms of a dialectic between nervous and reproductive systems. The stage of possibility is set by a prior strategic realm dealing with resource intake and outflow. For example, in recent accounts food-getting strategies logically have to function as the foundational, generative variables; the progeny are sex and mind (Wrangham, 1979). Sex is one of two preeminent biological issues—or, really, not sex but reproduction, since sex is something of a scandal from the point of view of rational processes of copy fidelity. Sex introduces too much difference, and so costly conflict, without making the benefits completely clear. But primates are stuck with it, the sad burden of evolutionary inertia. (At least the hominids got rid of estrus, although this modest reform has caused a great deal of scientific turmoil, producing some of the most bizarre contributions to the primate literature. For a serious and amusing summary, see Hrdy 1983.) We are told sexual politics would not exist had the early cells not fooled around and ended up in escalating asymmetry. Sexual politics is theorized fundamentally as the result of original difference. Big egg, little sperm, presto the dialectics of history and the sad facts of the dismal science of economics. "The investing sex becomes the limiting resource" (Hrdy, 1981, p. 22; Trivers, 1972; Williams, 1966). The integration of biology into the international economic system is a social fact at metaphoric, theoretic, and technical levels. This is not the result of bad, ideological science, but a powerful example of the specific historical production of natural-technical objects of knowledge. Power is productive.

The second preeminent issue is — but what do you call it? Brain, consciousness, strategy, mind? By the late-20th century, both sex and mind have been recast from organismic molds into technological-cybernetic ones: they have become coding/control problems for systems that are still nostalgically called *organisms*. But, simulations really have more status than organisms for a thoroughly high status biological theorist. Ask any serious sociobiologist. Another way to put it is that the referent is less sexy than the sign. Realism gives way to post-modernism in biology as well as literature and film (Haraway, 1985; Jameson, 1984). Guess which human gender does more high status simulating of both sex and mind. (Are you sure?) "Strategic reasoning" in several authors in primatology comes to be equated with rationality pure and simple. This is a wonderful origin story for the kind of reason that made writers in the Frankfurt school so nervous.* But, whether monkeys and apes are imagined as old-fashioned organisms or new-fangled coding systems, for primatology, sex and its control are inescapably what need to be known and explained.

Anthropology

Having rigorously demonstrated the importance of sex for biology, let me equally compellingly elucidate the structuring of possibility for a kind of knowledge called *anthropology*. Two things are crucial. First is anthropology's birth from the distinction between primitive and civilized, between nature and culture, between those who travel and look and those who stay home and are looked at. There is no way around the charged historical constitution of the Other as an object for appropriation, for observation, for visualization, for explanation. This structure has been generative of the sciences of man (sic). It works in art, politics, economics, science.

Primatology's other is doubly primitive, doubly the matter to the form of anthropology, because the object is really an animal. Or is it? The puzzle of primatology is precisely here. Does one do cultural anthropology of monkeys?

---

*A clear origin story privileging strategic reasoning is de Waal's (1982) *Chimpanzee Politics*. Langdon Winner (1980, 1983) argues that artifacts have politics and discusses the consequences of the reduction of public reasoning (i.e., politics) to questions of cost-benefit strategy. That strategy is only one way of resolving difference. My modest amendment to his argument is to note that animals have politics articulated in models of mind written into their pliant heads and genes (Haraway 1979, 1981–1982). The classic text is Dawkins (1976).

Sociobiologists accuse anthropological primatologists of doing little else. It is almost not a joke to imagine a truly dialogic relation with apes, in which experimental ethnography and coauthorship can be attempted (Clifford, 1983). Jokes are always about possible boundary incursions. Primatology is about the simultaneous and repetitive constitution and breakdown of the boundary between human and animal; that is, this aspect of primatology is about the moment of origins again and again. Primatology is a time machine in which the other is placed at the time of origins, even if the empirical field is in modern Rwanda or Kenya. It is not an accident that the objects of primatology live in the Third World; they are the preeminent tropical other, happily literally living in a vanishing garden.

However, by 1980, it is about as difficult to find a truly natural primate as a truly natural "savage"; decolonization makes naturalization very hard and subverts whole fields of knowledge. How do you have a proper National Geographic style field experience alone with the apes amidst a crowd of camera-clicking tourists bringing in needed foreign exchange? (Jane Teas, personal communication). Poachers are even less funny (Fossey, 1983). Field primatologists go to great lengths to structure the natural status of their objects of knowledge (Haraway, 1983b). For human anthropologists, the problem was the ethics of exchanging tobacco for the raw materials of textualization (Shostak, 1981). For primate anthropologists, the problem is whether to touch, how close to come if the other is to be wild, still the mediator of the passage at the time of origins.

Both human and primate anthropologys necessarily obliterate the other, at least as a natural other, in the process of textualizing it. The race is to write just ahead of extinction. (The issue is less literal extinction than preservation of the natural status of the object of knowledge. Primates and primitives are disappearing as proper dream objects for Western knowers.) One result is the inescapable immersion of primatologists in the politics of international conservation, a matter that interacts with and greatly complicates the politics of gender. Though not an anthropologist, National Geographic's Jane Goodall represents the perfect gendered condensation of these dilemmas: the lone, white, woman scientist mothering her blond son in the image of Old Flo, the perfect chimp mother, while deep into the night the human types her field notes that make Flo and her kind safe in books far from a Gombe penetrated by Zairian guerillas.

The question of touch, of closeness to the primate object of knowledge, has been mediated specifically by women primatologists. The National Geographic films on the women who have studied apes are constructed as extraordinary orgies of touch. From the quest for a first glimpse through the obscuring jungle to the mutual embrace with a trusting brown ape, the white women are written into a compelling, sensuous, filmic narrative of a need for "nature." The touch is explicitly named as a healing connection with "man's" animality, with origins, with a healthy union before the transgressions of the city and the bomb. One member of an ambiguous category can come closer to a member of another ambiguous category, and woman and animal are closer epistemically to each other within the tortuous logic of nature and culture than are man and animal (Ortner, 1972). Women primatologists have gone to great lengths to try to evade this polluting legacy (interviews with Adrienne Zihlman, Jeanne Altmann, Jane Teas, Shirley Strum, Naomi Bishop). They have also gone to some lengths to capitalize on it, turning dirt into gold, touch into science: natural symbols (Douglas, 1970).

Obviously, sex and gender cannot be avoided in life and human sciences. Western man needs sex. Equality (not to mention domination) requires the same of Western woman.* Sex and gender structure knowledge: they are the object of knowledge and the condition of knowing. This is my second crucial point about anthropology. Anthropologists have prided themselves for their early attention to sex and gender compared to other human or social sciences, but another face of this achievement is that this discipline, with psychology and biology, has been a leader in the constitution of sexualized discourses (Foucault, 1978). Is "the sex-gender system" a discovery of the first importance (Harding, 1983), or is it an over-determined position within the logic of nature–culture, full of the implicit problems of the latter pair? Both, obviously.

Many commentators have noticed the similarity of the categories: woman-animal-primitive-other-body-resource-child-matter-potency. They are all sinks for the injection of meaning. Is that what the privileged signifier is all about? Strathern (1980) argues that Western nature–culture polarities necessarily relate as resource to achieved production, matter to form, that which is appropriated to the active appropriator. She argues that because of this structure, the apparently innocent equation of nature–culture to Hagan (and implicitly other) distinctions like wild–domestic is actually a very serious, politically laden mistranslation. Bioanthropology is even more deeply infused with this problematic relationship than human anthropology. But it is only animals. . . .

---

*Everybody knows orgasms are highly political; that's why female monkeys had to have them in the last few years. It took some ingenuity to engineer them in the lab, but now observations are properly replicable in field and lab (Burton, 1971; Chevalier-Skolnikoff, 1974). Sarah Blaffer Hrdy bases *The Woman That Never Evolved* (1981, written partly in response to Symons, 1979) at least as much on Mary Jane Sherfey (1973) as on Darwin. Sherfey has more page references in Hrdy's index than E. O. Wilson (1975). Hrdy is a much better political theorist than Wilson.

Feminism

Feminist epistemology, political theory, and scientific discourse inherit the problems of humanism. If humanism in many of its forms constructs man through the logic of appropriation of primitive-civilized, Western feminism has constructed its object—along with the claim of recognizing male supremacy cross-culturally—through the logic of sex-gender. The very problematic "object" of feminism has been woman, but woman "under erasure" (see especially the journals, *Questions Feministes* 1977–1980 and *Feminist Issues* from 1980). But note that "in the beginning" feminist theory reconstructed its object, so that a feminist theory of woman is used as a lens to *see* the universal (or exceedingly common) domination of women, rather than the historic unity of man and his emerging project of self-realization in the activities of men. Alternatively, feminist theory uses a reconstructed version of woman to insist on the dispersion of the category, the irreducibility of the differences among women to any single category woman. Woman is plural here; the word granulates in one's mouth (Sandoval, n.d.). Among the possibilities opened by this strategy in theory is the longed-for discovery that women have *not* always been subject to male domination, that hope is also in the past, not only the future (Rosaldo, 1980). Also opened up is the knowledge that women too are practiced dominators.

So feminist theory creates a genealogy with the aid of the operator of sex-gender. A genealogy is an origin story; it assigns positions from which meaning flows. Primatology is inherently a genealogical practice in this sense. Within the field of primatology, all the possible positions for the meaning of being female can be and are being generated. Primatology is a field for contesting basic categories structured by the axis or net of sex-gender. The possible meanings of being female in the primate order are at the center of primatological discourse. The constantly ambiguous, equivocating objects and boundaries of primatology are made for this discourse. What is female to woman? female to females? woman to women? What is at stake, and for whom, in recording the intimate details of the lives of female and male baboons and chimpanzees?

Sex, the raw material of gender, remains a kind of generative resource, potentially free and freeing, but everywhere bound by the politics of gender. Potency has always seemed more innocent than act. The "end of gender" then becomes one possible feminist goal, a very tantalizing one. The very category "woman" is a scandal necessitating the end of the conditions that produced it. But if "gender" goes, so must "sex," just as the "West" cannot be held together without its "others" and vice versa. For feminist theory, a core problem has been the prior construction of woman/female within a hostile social and epistemic field structured by the nature-culture axis or net, as well as by the sex-gender axis. But although it is clear woman is the (political) gender of the (constructed) sex female, what are women? (Men also have an odd relation to man, not to mention to male. But it's not a random relation.) What is the object (both goal and subject matter) of feminism? What is the "natural-technical object of knowledge" of feminism considered as a "human" and a "natural" science? How does the object of feminist theory and practice relate to the "natural-technical *objects*," in both senses, of primatology?

Animals, especially the boundary animals, which primates are preeminently, serve as special objects for understanding the origins of socialized sex, almost gendered sex. This is serious business in the politics of domination and liberation, in which the source of energy, the self-replenishing luminescence outside the dead light of reason, must be located—and appropriated in another chapter in the story of domination. Culture and personality studies have turned regularly to the psychobiology of sex for good reason. The boundary between sex and gender, ever invisible but ever essential to visualize, must be sought. The boundary between sex and gender is the boundary between animal and human, a very potent optical illusion and technical achievement. Primatologists, including women primatologists, have focused extraordinary attention on sexual behavior. It readily carries the critical meanings in the origin of sociality explored through the logic of nature and culture.

Women primatologists have focused on female primates' sex (sexuality? Or, is that term reserved for humans?) partly to remove it from the inert, natural state it attained in the texts of their primatological brothers. The category female has been reconstructed in ways analogous to reconstructions of the category woman. Female sex was mere resource for male action that got animals to the border of humanity. But no more; female sex now has the promising dual properties, both active and natural, that let it serve as the mediator for the passage to culture as well. No longer just the tokens of male exchange, female primates have become sexual brokers in their own right. Female sex has become very active, social, and interesting—not to mention orgasmic across the primate order—in the last 15 years (Burton, 1971; Chevalier-Skolnikoff, 1974; Foucault, 1978; Lancaster, 1979). Solly Zuckerman (1932) would hardly recognize primate society. Female sex has been socialized and actualized, a critical move to make females political actors within the epistemic fields I am trying to characterize. One expected result is that females also get to have strategic reason to manage their investments wisely, a fine twist on "maternal thinking" (Altmann, 1980; Ruddick, 1982). Nice to be "informed."

The plethora of retold tales in the complex history of primate studies raises the central question of this essay: Is any other meaning possible for politics, the classic project to craft a public world from the chaos of difference, than war and domination, masked by logics of exchange (Hartsock, 1983a)? Have the simian lives narrated by women scientists really been different? Like that model of individuality and community, the slime mold, might primatology have some fruiting bodies rising from its hungry utopian project that would seed new meanings of power? Without prejudicing the reader's opinions on these authorial intentions before the end of the paper, let me suggest some mediate conclusions to guide reading.

First, although primatology is full of ideology in the old simple senses, it is dull and wrong to consider the matter of sexual politics addressed through unmasking ideology. There is no conspiracy of capitalist patriarchs in the sky to create a science of animal behavior to naturalize the fantasies of 20th century American white men, no matter how tempting the evidence sometimes seems. And struggles for a feminist science cannot proceed only by writing the tales one wants to be true, though "we" all do it. It is important not to trivialize the very real difficulty of good scientific storytelling. Gender and sex are central to the constitution of primatology, but in constantly complex ways and in interaction with multiple other interlacing, structuring axes that form the web of Western discourses. The very constitution of sex and gender as objects and conditions of knowledge—and so political categories—is at issue in feminism and in feminist readings/productions of primatology.

One of my informants concerning primates, a senior male scientist, argued that sex and gender *do* matter in primate science, determining how one knows, but that the "variable" is swamped by a host of others, obscuring differences between the sciences women and men craft. I am arguing he suggested the wrong metaphors. We are not looking at "variables," which could be ranged as dependent and independent and perhaps weighted through a savvy application of multivariate analysis, nor for essential differences between the practices of women and men as solutions to the key questions, though those differences are not trival. We are looking instead at the practical and theoretical constructions of a narrative field in which the explanatory model is better drawn from semiotics and hermeneutics than from statistics (Semiotics: the sciences of making meanings. Hermeneutics: the sciences of interpretation. Statistics: the sciences of arranging numbers.). But I hope for a politicized semiotics, where politics is the search for a public world through many socially grounded practices, including primatology. How could primatology not be a territory of feminist struggle? Western women's place is indeed in the jungle. Whether other women and men occupy that material/mythic space when they watch monkeys and apes is a function of other histories and other stories.

THE JUNGLE: SCENES

In moderation, numbers never grounded the flight of interpretation. About how many women practice primatology for a living? The question is difficult to answer for many reasons. My focus is on field primatology, that is, studies of wild or semifree ranging but provisioned animals in an environment that can be epistemically constructed to be "natural," a possible scene of evolutionary origins. But primatology is both a laboratory science and a field science that crosses dozens of disciplinary boundaries in zoology, ecology, anthropology, psychology, parasitology, biomedical research, psychiatry, conservation, demography, and so on. Primatologists from the United States are likely to belong to three major professional associations, but many individuals who made major contributions and who allowed me to interview them and read their unpublished papers do not appear on the membership lists ever or for several years at a time.

Making many assumptions, I will use 1980 membership lists from the American Association of Physical Anthropology (AAPA), the American Society of Primatologists (ASP), and the International Primatological Society (IPS) to suggest the present level of participation of women in field primatology. These global disciplinary counts present a minimum picture, because there is good reason to believe women are more heavily represented in field primatology than in exclusively laboratory-based practices, and they have been more authoritative in field primatology whatever their numbers. There is no absolute division between field and lab, but there is a tense difference of emphasis, despite the official doctrine that naturalistic studies require complementary laboratory studies with their greater power of experimental manipulation.

I am ignoring the large issue of skewed emphasis by concentrating on North Americans, with nods to the British, despite the important fact that primatology emerged as an international passion in the late 1950s. With a few important exceptions, the authoritative spokeswomen and the largest numbers of women in primatology have been U.S. nationals, trained and/or employed in U.S. institutions. The overwhelming majority, relatively more than for other biological sciences, have been white, although that is now changing.

In 1977–1978, the IPS (founded in 1966) roster listed 751 members, of whom 382 listed U.S. addresses, 92 U.K. addresses, 115 Japanese, 14 African (10 from South Africa), and 151 from other locations. In the IPS, overall, women were 20% of the membership: 22% of the U.S. total, 22% of the British, 9% of the Japanese, and 24% of the "other." Many individuals could not be identified by gender from initials; they were left out of these calculations, probably resulting in understating the representation of women. In general, the field staffs of permanent field research sites do not show up in professional rosters; this results in making the production of field primatology appear a more white affair than it is. As skilled staff are increasingly nationals of the countries where nonhuman primates live, this is misleading; but written primatology, outside the reports of national parks and internal documents from the research sites, is overwhelmingly authored by people like the professionals on these primatological society lists. The lists are also biased toward doctorate scientists. But for signs of change, see Baranga (1978) and Goodall et al. (1979). By subdiscipline, women accounted for 22% of the anthropologists, 12% of the medical researchers, 27% of the psychologists, 19% of those involved primarily with zoos or wildlife conservation, 19% of the zoologists or ecologists, 25% in other categories. (Percentages of total membership that could be ascribed to these subdisciplines are 17% for anthropology, 20% for medicine, 16% for psychology, 3% in zoos and wildlife, 9% of the zoologists or ecologists, and 25% for other.) In a rough way, women are relatively overrepresented (if that word can make sense when the level never equals 30%!) in anthropology, psychology, and "other," and are in zoos and wildlife and zoology and ecology in numbers about proportional to their presence in the society. Women are underrepresented in medicine. The very low number of Japanese women, despite the prominence of Japanese primatology internationally, is echoed in the difficulty of finding out much (using English-language sources) about them as individual researchers. This is in dismaying contrast to the prominence of Western women in the field.

The 1979 roster of the AAPA (founded in 1918) lists 1,200 persons, about 26% of whom appear to be women. Physical anthropologists have traditionally taught in medical schools and sought positions in museums, but that generalization is weak by 1980. The 1980 roster of the ASP (founded in 1966) lists 445 individuals, of whom only 23 are foreign, largely Canadian. About 30% of the ASP are women, including 45% of those who give themselves an anthropology-related address. Such an address reflects the fact that academic jobs for primatologists, whatever their discipline of training, are often in anthropology departments. About 24% of the women are psychologists, 36% of those in zoos or wildlife conservation, 20% of the total in zoology/ecology, and 47% of those whose interests intersect with psychiatry (compared to 11% in the IPS). Women primatologists appear to be trained in and/or have jobs in anthropology in proportions considerably higher than their representation in the association as a whole. The reverse appears to be true for zoology/ecology. Note that U.S. women primatologists appear to be more likely to join the American society than the international association, compared to U.S. men.

For comparison, the January 1982 National Science Foundation publication *Women and Minorities in Science and Engineering* notes that by 1978 in the United States women represented about 20% of employed social, life, and mathematical scientists, but only 9.4% of all employed scientists and engineers. (Contrast that with women's figure of 43% of all professional and related workers, disregarding stratification in what counts as professional, let alone "related.") About 85% of growth in employment of women doctoral scientists from 1973–1978 was in life sciences, social sciences, and psychology. Together life sciences (30%), social sciences (17.2%), and psychology (14.7%) account for 61.9% of women scientists. This is the pool from which primatologists come, and they come in numbers roughly characteristic to other life and social sciences. Nowhere does the representation of women equal 30%

of these global field listings. Except in psychology, in no category of sciences does the representation of women doctorates equal 20%. In the face of these unspectacular showings, women primatologists stand out slightly. Their impact has been greater than their numbers, compared to most other areas of anthropology and all other areas of biology. For this conclusion, I turn to their practice and their publications.

Field primatology is a recent undertaking, where almost all work has been done since the late 1950s; the period since 1975 represents the steepest growth of primate field studies. The explosive growth of primatology has overlapped the "second wave" of the Euro-American women's movements. Young women and men entering primatology in those years could not be unaware that their field was contested from the "outside," in gender politics and much else. It was also contested from the "inside." One result was the explosion of writing by women on primate society and behavior, both for popular and professional readers. The following lists, consisting only of books, hardly the major form of publishing, especially in natural sciences, is not exhaustive; but it gives the flavor of abundance and a chronology showing the steady rise in women's production in primatology.

Nadie Kohts' *Untersuchungen uber die erkentniss Fahigkeiten des Schimpansen aus dem zoopsychologischen Laboratorium des Museum Darwinianum in Moskau* (Moscow, 1923) opens my list in order immediately to transgress the categories of American, post-World War II, and field primatology, and also just to honor an important predecessor in the appreciation of primate mind. I include the next entry to mark the frequent role of the officially nonscientist wife, who contributes substantially to the production of the primate text: Robert Yerkes and Ada Yerkes, *The Great Apes* (New Haven, CT: Yale University Press, 1929).* The next entry is probably also little known except to the aficionados of apes, but the author marks several categories important to gender in primatology (zoo work, lay status, success in ape breeding): Belle Benchley, *My Friends the Apes* (Boston: Little, 1942). These three pre-World War II names are also included to underline my inability to find a single book, popular or professional, about primates by a woman PhD scientist in the world before the 1960s. There are several by men.

Then comes the best known name of all, beginning a chronological list of professional biologists and anthropologists writing for many audiences: Jane van Lawick Goodall, "My Friends the Wild Chimpanzees" (*National Geographic*, 1967), followed by *In the Shadow of Man* (Boston: Houghton Miff-

*Beginning in 1929, but with almost all entries since 1965, I have counted about 65 married couples publishing together in primatology, including a cursory count of laboratory psychologists and a more careful count of field workers. My count misses most couples outside England and the United States. This figure is a significant proportion of all active primatologists.

lin, 1971); Thelma Rowell, *Social Behaviour of Monkeys* (Baltimore, MD: Penguin, 1972); Alison Jolly, *Lemur Behavior* (Chicago: The University of Chicago Press, 1966), *The Evolution of Primate Behavior* (New York: Macmillan, 1972); Jane Lancaster, *Primate Behavior and the Emergence of Human Culture* (New York: Holt, Rinehart and Winston, 1975); Sarah Blaffer Hrdy, *Langurs of Abu* (Cambridge: Harvard University Press, 1977) and *The Woman That Never Evolved* (Cambridge: Harvard University Press, 1981); Alison Richard, *Behavioral Variation* (Lewisburg, PA: Bucknell, 1978) and *Primates in Nature* (forthcoming); Jeanne Altmann, *Baboon Mothers and Infants* (Cambridge: Harvard University Press, 1980); Katie Milton, *The Foraging Strategies of Howler Monkeys* (New York: Columbia University Press, 1980); Nancy Tanner, *On Becoming Human* (London: Cambridge University Press, 1981); Linda Marie Fedigan, *Primate Paradigms* (Montreal: Eden Press, 1982); Adrienne Zihlman, *Human Evolution Coloring Book* (New York: Barnes and Noble, 1982); Dian Fossey, *Gorillas in the Mist* (Boston: Houghton Mifflin, 1983). Several other books are in progress.

A larger picture emerges if we consider the profusion of books focused on debates about sex and gender that take serious account of the work of women primatologists and reconstructed men primatologists. Every one of these books is part of a large international social struggle, especially since the 1960s, about the political–symbolic–social structure, history (natural and otherwise), and future of woman/women. The political struggles are not context to the written texts. The women's movements, for example, are not the "outside" to some other "inside." The written texts are part of the political struggle, but a struggle conducted by very specific "scientific" means, including possible stories in the narrative field of primatology. By definition, the origin point has to be outside the history I will tell, therefore consider first the unique, renegade pre-1960s classic, a book that is to female primates and feminist primatology as Simone de Beauvoir's *Second Sex* is to feminist theory of the second wave: Ruth Hershberger, *Adam's Rib* (New York: Pellegrini and Cudahy, 1948, reissued in paper by Harper and Row, 1970; hardly an accidental date). Hershberger's dedication of the book to G. E. H. (G. Evelyn Hutchinson), a major scientist who made a habit of supporting heterodox women scientists, also marks the crucial importance of pro-feminist men in the prehistory of feminist struggles for science.

A title from the 1960s gives the starting point for thinking about females with regard to (zoological) class, but note how the field expands through the 1970s, when maternal behavior is no longer the totally constraining definition of what it means to be female: Harriet Rheingold (Ed.), *Maternal Behavior in Mammals* (New York: Wiley, 1963); Elaine Morgan, *Descent of Woman* (New York: Stein and Day, 1972); Carol Tavris (Ed.), *The Female Experience* (Delmar, CA: Communications/Research/Machines, 1973); Rayna Rapp Reiter (Ed.), *Toward an Anthropology of Women* (New York:

Monthly Review Press, 1974), with the "classic" paper by Sally Linton, "Woman the gatherer: male bias in anthropology"; Evelyn Reed, *Woman's Evolution* (New York: Pathfinder, 1975); M. Kay Martin and Barbara Voorhies, *Female of the Species* (New York: Columbia University Press, 1975; dedicated to Margaret Mead); Ruby Rohrlich Leavitt, *Peaceable Primates and Gentle People* (New York: Harper and Row, 1975); Cynthia Moss, *Portraits in the Wild* (Boston: Houghton Mifflin, 1975); Bettyann Kevles, *Watching the Wild Apes* (New York: Dutton, 1976); H. Katchadourian (Ed.), *Human Sexuality: A Comparative and Developmental Perspective* (Los Angeles: University of California Press, 1978); Lila Leibowitz, *Females, and Families: A Biosocial Approach* (Belmont, CA: Duxbury, 1978); Lionel Tiger and Heather Fowler (Eds.), *Female Hierarchies* (Chicago: Beresford, 1978); W. Miller and L. Newman (Eds.), *The First Child and Family Formation* (Chapel Hill, NC: Carolina Population Center Publications, University of North Carolina, 1978); Elizabeth Fisher, *Woman's Creation: Sexual Evolution and the Shaping of Society* (New York: McGraw-Hill, 1979); Frances Dahlberg (Ed.), *Woman the Gatherer* (New Haven, CT: Yale University Press, 1981); Helen Fisher, *The Sex Contract: The Evolution of Human Behavior* (New York: Morrow, 1982); Ruth Bleier, *Science and Gender: A Critique of Biology and Its Theories on Women* (New York: Pergamon, 1984). It would be a serious mistake to leave out science fiction, which is both influenced by and an influence on the struggles over sex and gender in primatology; for example, Jean Auel's *Clan of the Cave Bear*; Marge Piercy's *Woman on the Edge of Time*; and the bio-fiction of C. J. Cherryh and James Tiptree, Jr., both women science fiction writers despite nominal appearances.

This list is heterogeneous from several points of view: political allegiance, intended audience, credentials of the authors and editors, publishing format, genre, etc. Interestingly, it is nationally and racially homogeneous; this point matters in view of the universalizing tendency of the literature, which repeatedly seeks to be about the nature of "woman." No one could claim from any of the lists in this paper that white U.S. women occupy a unified ideology or are in any simple sense "in opposition" to masculinist positions, much less to men. But it should also be impossible to miss the collective impact of these public, ordered stories: new lines of force are present in the primate field. It has become impossible to hear the same silences in any text. The narrative field has been restructured by a polyphony rising from alalia to heteroglossia. In the practice of telling important stories of origin among peoples of the Book, women now also speak in tongues, imagining female within a native language (Elgin 1984).

Volumes edited or coedited by professional women primatologists produce another long list that begins with the publication of the papers from Phyllis Jay's 1965 Wenner Gren conference (Jay, 1968). The list ends for now with a spate of books published in the mid-1980s. These books mark the newly hegemonic place of "sociobiological theory" in primate studies and the complex place of sociobiology in the crafting of self-consciously pro-female and often feminist accounts of primate, and indeed vertebrate, evolution, behavior, and ecology (Hausfater & Hrdy, 1984; Small, 1984; Wasser, 1983). Meredith Small's *Female Primates: Studies by Women Primatologists* is explicitly a celebration of female primates, human and animal, in collaboration to write primatology. It is also published as volume 4 of *Monographs in Primatology* under the scrutiny of a nine-member editorial board, only one of whom (Jeanne Altmann) is a woman. The editor is a graduate student and she was explicitly encouraged by her male advisor, Peter Rodman. *Female Primates* includes 21 women and 1 man (as a coauthor) among its authors. The range of concerns includes postmenopausal animals, female adolescence, female sexual exuberance, feeding strategies, mating systems explained from the point of view of female biology as the independent variable, and much else. Any notion that the book might be pollutingly popular should be nipped by a combination of style and a $58 price tag; it is professional to the core.

Because it is a kind of summing up and celebration of primate females and the women who made them visible (i.e., a construction of a "we"), *Female Primates* deserves a full analysis. For now, however, I will only look briefly at two pieces for their strategy in introducing subsequent papers and thus framing the whole enterprise. Each piece raises the question of whether it makes a difference that women primatologists focus on female animals, but each also adopts a philosophy of science and ideology of progressive improvement of knowledge that block an investigation of an epistemic field structured by sex and gender. From the point of view of the framing pieces, "male bias" exists but can be corrected fairly simply. There is no need for dangerously political social relations within primatology and no need for the matter to challenge the practitioners' "native" account of how knowledge is made, at least not in public. Bias cancels bias; cumulative knowledge emerges. The root reasons given, however, hint at a stronger position: only bias ("empathy") permits certain "real" phenomena to be knowable, or only explanation from the point of view of one group, not the point of view of an illusory whole that actually masks an interested part, gets at the "real" world. In this case, bias or point of view turn out to be the social and epistemic operator, sex–gender. The major scientific–political question is how such a potent point of view is constructed. In the construction of the female animal, the primatologist is also reconstructed, given a new genealogy. But the rebirth is within the boundaries of the West, within its ubiquitous web of nature–culture. Primatology is simian orientalism (Said, 1978).

Jane Lancaster, who introduced the volume as a whole, was a PhD student of Sherwood Washburn at the University of California at Berkeley in 1967 and a senior student of primate behavior from anthropological points of view. The introduction is remarkable for its adherence to sociobiological and socioecological perspectives; it is a sign of the triumphant status of those explanatory frameworks in evolutionary biology, including primatology by the mid-1980s. Within that frame, Lancaster looks at primate field studies to understand four areas of sexual dimorphism: "sex differences in dominance, mating behavior, and sexual assertiveness, attachment to home range and the natal group, and the ecological and social correlates of sex difference and body size" (Lancaster in Small, 1984, pp. 7–8). In each case, the point is that "females too do x." It turns out that (a) females are competitive and take dominance seriously; (b) females also wander and are not embodiments of social attachment and conservatism; (c) females also are sexually assertive; and (d) females have energy demands in their lives as great as those of males. Focus is on females and not on

> the species as evolving as an amorphous whole. We explore the social world of females rather than that of the social group. . . . We learn to understand the reproductive strategies of females and to balance these strategies against those pursued by males of their social systems. . . . At last we are coming to a point of balance where the behaviors and adaptations of the sexes are equally weighted. (Lancaster in Small 1984, p. 8)

Finding females means disrupting a previous whole, now called *amorphous*, rather than the achieved potential of the species. Feminism absolutely requires breaking up some versions of a "we" and constructing others.

Lancaster's is a very interesting construction of a "we" where the boundary between female animal and woman primatologist is blurred, ambiguous. The deliberately ambiguous title of the whole volume is echoed again and again: "we" are all female primates here, outside of history in the original garden. That garden naturally turns out to be in the liberal West. Competition, mobility, sexuality, and energy: these are the marks of individuality, of value, of first or primate citizenship. "Balance" is equality in these matters, hard won from specific attention to the point of view not of the "amorphous whole" but of "the social world of females." Lancaster's is an origin story about property in the individual body; it is a classic entry in the large text of liberal political theory, rewritten in the language of reproductive strategy. Sex and mind again are mutually determining. In the reconstruction of the female primate as an active generator of primate society through active sexuality, physical mobility, energetic demands on self and environment, and social competition, the woman "primatologist," that is, female (human) nature, is reconstructed to have the capacity to be a citizen, a member of a public "we," one who constructs public knowledge, a scientist. Science is very sexy, a question of eros and power. Appropriately, this "we" is born in an origin story, a time machine for beginning history, therefore outside history.

Thelma Rowell, a senior zoologist at the University of California at Berkeley who played a major role in disrupting stories about primate social behavior, especially stories about dominance (Rowell, 1974), was invited to introduce the first subcollection of papers, called "Mothers, Infants, and Adolescents." Rowell's message as always was about complexity. She is not hesitant to point out the legacy of male bias in primatology, for example, in the classification of females as juvenile or adult exclusively as a function of their capacity to breed, while males were categorized by a whole series of stages grounded in social as well as minimal reproductive functions.

> For that matter, there is little recognition of continued social development in human females, which for most purposes are also classified as either juvenile or old enough to breed. In contrast, continued social development following puberty in males was recognized in the earliest studies of primate social behavior, just as the stages of seniority are often formally recognized among men. This dual standard has, I think, delayed our understanding of primate social organization. (Rowell in Small, 1984, p. 14)

She points out the merit of the following papers in seeing the primate world from the "female monkey's point of view" and, thereby, "challenging accepted explanations." She goes further, writing, "I have a feeling it is easier for females to empathize with females, and that empathy is a covertly accepted aspect of primate studies—because it produces results" (Rowell in Small, 1984, p. 16).

But she backs off from exploring unsettling implications of these positions about the structuring of the observer determining the possibility of seeing. Instead, because males identify with males and females with females and primatology attracts both human genders, the result is additive, canceling out "bias" and leading toward cumulative progress: "The resulting stereoscopic picture of social behavior of primates is more sophisticated than that current for other groups [of mammals]" (p. 16). But the stories are not stereoscopic, where the images from separated eyes are interpreted by a higher nervous center; they are disruptive and restructuring of fields of knowledge and practice. The reader is not an optic tectum, but a party to the fray, so hope for higher integration from that source is futile.

Further, "empathy" produces results in human anthropology as well, forming part of a very mixed legacy that includes universalizing, identification, and denial of difference, as the "other" is appropriated to the explanatory strategy of the writer. Empathy is part of the Western scientific tool kit, kept in constant productive tension with its twin, objectivity. Empathy is coded dark, covert, or implicit, and objectivity light, acknowledged, or explicit. But each constructs the other in the history of modern Western science, just as nature–culture and woman–man are mutually constructed in a logic of appropriation and progress. When Lancaster wants to see "balance" and Rowell writes about a "stereoscopic picture," they simultaneously raise and dismiss the messy matter of scientific constructions of sex and gender as objects of knowledge and as conditions of knowing. Official (or native) philosophies of science among researchers obscure the complexity of their practice and the politics of "our" knowledge.

It is arguable that the highest status science is coded as requiring the greatest empathic and intuitive capacities, exhibited in a special way by the representatives of the gender man, not woman. Kekule's dream of the benzene ring is an example; whole parts of the chemical industry rest on that night. Einstein, Polanyi, Chargaff, Faraday, other physical scientists, especially those located in theoretical physics and mathematics, are ascribed special abilities to intuit the world. Ascription of genius does not rest on the ideology of objectivity alone, or even principally. However, not surprisingly, the same sciences are coded as exacting the greatest powers of rational discrimination and "objectivity." Gender coding is necessarily contradictory, or it could not be the powerful operator that it clearly is. Everyone would slip through the net, and unfortunately none of us does, although we do successfully degender parts of ourselves from time to time. It's a bit risky, even for the privileged. Nuanced and contrasting consideration of these issues from feminist points of view is found in Traweek (1982) and Keller (1983, 1985).

The portrait of publishing and rough numerical representation needs to be complemented by a brief survey of the major institutions that have produced women scientists in the field. Women's professional practice in field primatology has meant access, submission, and contribution to the institutional means of producing knowledge. Despite the *National Geographic*'s imagery of Jane alone in the jungle with the apes, a PhD is bestowed for *social* work, often experienced as lonely and sometimes named as alienated, in a different sort of jungle where monkeys and apes are transcribed into texts or, more recently, coded onto tape.

Even Tarzan learned to read; he is in fact one in a long line of autodidacts, the issue of bibliogenesis. Progeny also include Frankenstein's monster; Tarzan's author, Edgar Rice Burroughs; and Frankenstein's author, Mary Shelley. *National Geographic* films of Jane Goodall show her alone deep into the night transcribing her field notes; the day is recording, the night is transcribing. The filmic text is about the hope of touching nature and being accepted. The films hardly hint at Gombe's elaborate history of record keeping, involving dozens of workers from many countries over 25 years, with the aid of major universities (such as Stanford, the University of Dar Es Salaam, and Cambridge) and the assistance of capacious computers, tape recorders, and other paraphernalia of modern science writing.

Women did not earn PhDs for research on primate behavior randomly from all possible doctorate-granting institutions where people did such work. For example, Harry Harlow's laboratories at the University of Wisconsin were particularly scarce of human female doctoral fauna, a fact that contributes to the pattern of more women in the field than in lab-based psychological primatology. Two universities were initially crucial, the Anthropology Department of the University of California at Berkeley and the sub-Department of Animal Behavior of Cambridge University at Madingley. By the 1970s, Stanford University's program in human biology, with ties to Gombe and its captive chimpanzee colony, and Harvard University's program in physical anthropology became important from the points of view of this paper.

My counts are not final. From Sherwood Washburn's initiation of the seminar on the "Origins of Human Behavior" at the University of Chicago in 1957–1958 and his move to Berkeley in 1958, with the establishment of an animal behavior experimental station and field studies of primates all over the world, until his retirement from the University of California in 1980, at least 18 women earned PhDs for work on primate evolution and behavior in a program deeply influenced by his plans for reconstructing physical anthropology and the explanations of human evolution. Many of those women were the students of Washburn's former student, Phyllis (Jay) Dolhinow, who came to Berkeley in 1966. This program has been famous for its unusually large number of women students in the early years of post-war primatology. The role of Washburn in the accomplishments of his students is controversial and many other figures were crucial to their intellectual formation: for example, Peter Marler, Frank Beach, and Thelma Rowell. But the program founded and sustained by Washburn's power in physical anthropology was the route to credentials for the U.S. women until the late 1970s, as well as most of the men through the 1960s.

Many of the UCB women have been leaders in reconstructions of sex and gender in primate storytelling. They provided peer cohorts for each other during graduate school and formed critical support networks in later professional life. Their relationships with their men student peers are an important part of the story. There are several "generations" of UCB women primatologists, not to mention individual heterogeneity, and generalizations are tricky. Their strengths and limitations are controversial and germane to the debates about explanatory powers of sociobiology and socioecology compared to evolutionary structural functionalism. The academic entrepreneurship of Washburn mattered enormously to the professional status and opportunities of these women and men. A fruitful way to follow their collective fates is by tracing the conferences funded by the Wenner-Gren Foundation from the high point of the Washburn network's influence in the early 1960s to the ascendancy of sociobiology/socioecology; that is, from the 1958 Darwin Centennial at the University of Chicago organized by Sol Tax and the 1965 follow-up conference, called inescapably the "Origin of Man," to the 1982 conference on "Infanticide in Animals and Man."

Like Washburn, Robert Hinde of Cambridge sponsored the doctoral work of a significant number of the important women primatologists, including Jane Goodall and Dian Fossey of *National Geographic* fame. At least as important has been the work of Thelma Rowell, an early Hinde student, who, after several years at Makerere University in Uganda, moved to the Zoology Department at Berkeley, where her presence made a major difference to the students of primatology in the Anthropology Department as well, perhaps especially the women. My interview informants have argued that Goodall and

Rowell were critical to Hinde's theoretical and methodological development, leading him to see beyond Lorenz and Tinbergen to the complexity and individuality of primate behavior. Including PhD students and postdoctoral associates, since 1959 about 15–20 women primatologists have been associated with Hinde's laboratory at Madingley. The approach of his lab may be followed in a recent collected volume (Hinde, 1983). Many of these students were Americans who earned their PhDs in the United States and did postdoctoral work with Madingley or vice versa. Networks of institutions and researchers are probably a more useful way to trace primate lineages than dissertation advisors. Crucial in these networks are the long-term field sites, like Gombe, Amboseli, Gilgil, Cayo Santiago, and a few others. Among the Gombe field workers, at least 35 have been women, including nondoctoral research assistants.

Stanford University was for a time at a nodal point of institutions and field sites, connected especially to Berkeley and Gombe. The entrepreneurship of David Hamburg was crucial. Hamburg and Washburn linked worlds in the year-long primate meeting at the Stanford Center for Advanced Study in the Behavioral Sciences in 1962–1963, resulting in one of the first volumes on modern primate studies (DeVore, 1965). Hamburg was responsible for Stanford's fruitful collaboration with Jane Goodall and primate research at the Gombe Stream National Park in Tanzania, a collaboration that broke up tragically with the kidnapping of Stanford graduate students at Gombe in 1975. But during the Hamburg–Goodall years, several students of primatology were formed, including many women whose networks have been part of the restructuring of primatology since the challenges of sociobiology of the mid-1970s. Stanford women, former undergraduates as well as graduate students, have important ties with other central institutions grounding primate research. In addition to Gombe, they worked at Harvard, Cambridge, University of California at Davis, Kekopey Ranch at Gilgil, Amboseli, the University of Chicago, the Rockefeller University's research station at Millbrook, and other places. Their ties with each other and male peers were crucial to setting up a second primate year at the Center for Advanced Study in the Behavioral Sciences in 1983–1984, to produce another volume reflecting the recently ascendant explanatory frameworks. It is certain that reconstructed female animals, as well as women primatologists, will occupy very active positions in that text.*

*Dorothy Cheney, Robert Seyfarth, Barbara Smuts, Thomas Strusaker, Richard Wrangham (forthcoming). In general, the Berkeley women are in quite different networks than the Stanford–Harvard–Cambridge webs. Partly, the difference is the cleavage between zoological–ethological and anthropological frames. Those who cross the cleavage are particularly interesting, but, in general, the traffic on the bridge is mostly in the direction of adopting the sociobiological–socioecological strategies. From another point of view, the webs among younger workers, especially the women, simply do not follow the cleavages set up by the famous controversies.

Irven DeVore has been the dominant figure in Harvard's program in physical anthropology since he finished his PhD in 1962, from many accounts as Washburn's favored son. DeVore's early baboon field study was a central leg of the man-the-hunter research program, and the male orientation of that baboon report has been notorious. (It has also been the standard source for school textbooks and the TimeLife series on animals. The man-the-hunter program was "tri-pedal," with legs in functional anatomy, primate field studies, and anthropological investigation of human hunter-gatherers. Richard Lee, also from the Washburn Berkeley world, partly in collaboration with DeVore, has been of fundamental importance here. Lee's pro-feminist reputation and publications contrast markedly with DeVore's.) DeVore's undergraduate course in primate behavior at Harvard has been immensely popular and, since DeVore's famous "conversion" to sociobiology (to Washburn's great dismay) in the 1970s, that course and his graduate primate seminar have been important institutional mechanisms for reproducing the explanatory strategy in younger workers. Robert Trivers' tutelage of DeVore has been a central feature of the framework. It also appears that in the first sociobiological years the seminars were classically "male-dominated," by faculty and students.

But then the name of Sarah Blaffer Hrdy begins to appear in print and in my informants' accounts. An unrepentant sociobiologist, she has centered females in her accounts in ways that have destabilized generalizations about what "sociobiology" must say about female animals or human women. She is also an unrepentant feminist, greatly admired by the reviewer of her *Woman That Never Evolved* in *Off Our Backs*, the major national radical feminist newsprint publication in the United States, and greatly criticized by socialist feminist opponents of liberal political theory, including its sociobiological variants. I have been in the latter camp, but fortunately Hrdy is not so easily bundled off. She is considerably more complex than the labels imply. Hrdy is controversial on several accounts, from how she is perceived in relation to other women, to the politics and science of her field work and writing. In the present context, her role in the Harvard primate seminars is at issue. Women students who came to Harvard for graduate work after Hrdy consistently name her presence as a crucial supportive factor in their own confidence and intellectual power. They formed cohorts with each other and regarded Hrdy as an elder sister. These networks ground much of the currently interesting reconstructions of primate females and primate society as a disrupted "whole."

One last locus should be characterized: the savannah baboon research project in Amboseli National Park in Kenya and the Department of Biology (Allee Laboratory of Animal Behavior) at the University of Chicago, where Jeanne Altmann and Stuart Altmann have worked since 1970. The pattern of the married couple in primatology has been an important one, with the husband regularly better known. In some ways, the Altmann picture was similar, but there are refreshing differences that matter in the reconstruction of primatology. Jeanne Altmann has been important in primate field studies since she began working with Stuart Altmann in the early 1960s, but she earned her PhD only in 1979, with a dissertation ("Ecology of Motherhood and Early Infancy") submitted to the University of Chicago Committee on Human Development. The dissertation was a version of her important book (1980). In 1974, Jeanne Altmann published one of the most cited papers in field primatology, "Observational Study of Behavior: Sampling Methods." The simple title belies the importance of the paper in setting standards for nonexperimental research design, especially if the observer has any hope of doing reliable statistical analyses. Initially, J. Altmann, without a PhD, was rarely invited to conferences unless her husband was also invited. Progressively, she became a power in the field in her own right. Jeanne Altmann is cited by my younger informants as a significant node in developing "invisible colleges" among women. Her importance is loosely analogous to Lillian Gilbreth's in the history of personnel management research in the early years of labor studies right after the triumph of Taylorism. Gilbreth was a major theorist in this area of capitalist science. Jeanne Altmann is a theorist of the ergonomics of baboon motherhood, where ergonomics is a kind of cybernetics of the division of labor and a crucial concept in the deep construction of nature as a problem in investment strategy (Trescott, 1984).

THE TEXT: REPRESENTATIONS

Stories of the nature and possibility of citizenship and politics in Western traditions regularly turn on versions of the origin of "the family." The stage has been set, so let us conclude with deliberately parodic, humorous tales by Adrienne Zihlman and Sarah Hrdy, two bioanthropologists otherwise committed to very different storytelling strategies.

Both Hrdy and Zihlman are ardent feminists, convinced that such commitments are integral to their practice of *good* science. Their meanings of feminism and their ways of practicing science are in sharp conflict, but they both take "stories" seriously as part of their craft, not spare-time pursuits. Their task, within the contested constraints of discourses structured by nature–culture and sex–gender, is to give an evolutionary account of the human place in nature and society. Their best writing displays a complex reflexiveness about their own ideologies that emerges through conscious oppositional practice within simultaneously privileged and oppressive contexts. Hrdy and Zihlman are both "Milton's daughters." Neither had the luxury of professional formation in a symbolic culture and historical society whose stories and sciences were friendly to them and their kind. They inherited another status. Nonetheless, they had access to the resources of culturally authoritative storytelling: PhDs from major science institutions, significant financial resources, and the intellectual and emotional riches of a worldwide feminist resurgence coinciding with critical periods of their professional and personal formation. As Milton's daughters, these women scientists once read "to" the blind father, but they also read the book of nature for their own purposes. Zihlman retold the inherited stories of "man the hunter," beginning within the constraints of structural-functional physical anthropology. Hrdy recast the plots and characters of sociobiology, turning unpromising material into scientific and ideological resource.

The tales considered here were told in response to interpretations of the recent reappearance in the paleoanthropological field in Haadar, Ethiopia, of a diminutive, ancient (say, 3-million-year-old) hominid grandmother—of erect and bipedal habit, but small mind—named by her Adamic founders, Lucy, after the drug culture that gave their generation of students historical identity (Johanson & Edey, 1981). (The reference was to "Lucy in the Sky with Diamonds." Lucy could be Lucien, but let's give her her sex, since it is crucial to the story at hand. The paucity of African names in paleoanthropological and primatological literature says a fair amount about the limitations of Adam's claim to species fatherhood.) Lucy's near-complete skeleton was dug out of the earth by the skilled hands of a brotherhood, which recognized in her and associated skeletons a resource for re-establishing potent masculinist versions of the origin of man (Lovejoy, 1981). Lucy was quickly made into a hominid mother and faithful wife, a more efficient reproducing machine than her apish sisters and a reliable, if poorly upholstered, sex doll. These are the qualities essential to the male-dominant, "monogamous," heterosexual family, named "the family" with mind-numbing regularity. Lucy's bones were incorporated into a scientific fetish-fantasy, dubbed irreverently the "love and joy" hypothesis in Sarah Hrdy's response (Hrdy & Bennett, 1981, p. 7). But women still "dub", while men "name."

What makes Lovejoy's interpretations of Lucy "masculinist," as opposed to simply distasteful and controversial for his scientific opponents? The answer is his unwitting discipleship to the father of biology, Aristotle. Lovejoy's "Origin of Man" is enmeshed in the narrative of active, potent, dynamic, self-realizing manhood achieving humanity through reproductive politics: paternity is the key to humanity. And paternity is a world historical achievement. Maternity is inherently conservative and requires husbanding to become truly fruitful, to move from animal to human. Standard in Western masculinist accounts, *disconnection* from the category "nature" is essential to man's natural place: human self-realization (transcendence, culture) requires it. Here is the node where nature–culture and sex–gender intersect.

Lovejoy argues that the transition to a savannah-mosaic environment at the temporal boundary (late Miocene) marking hominization placed pre-hominids in a reproductive crisis requiring either closer birth spacing or greater survival of offspring or both. Expulsion from the forest garden meant a reproductive burden of species-making proportions. The narrative of matrifocal, female-centered worlds of apes had to give way to the more dynamic "human" family.

> In the proposed hominid reproductive strategy, the process of pair bonding would not only lead to direct involvement of males in the survivorship of offspring[;] in primates as intelligent as extant hominoids, it would establish paternity, and thus lead to a gradual replacement of the matrifocal group by a "bifocal" one—the primitive nuclear family. (Lovejoy, 1981, pp. 347–348)

Anthropologist Carol Delaney (1985) pointed out that paternity, in the hoary disputes in her discipline about whether real human peoples ever lived who *really* did not know about it, does not mean simply knowledge of a male biological contribution in conception. In Western patriarchal culture, it means what Aristotle meant: male reproductive causality in the medium of the receptive female. The blindness induced by masculinist privilege in the culture of the anthropologist made their own specific meaning of paternity opaque to them; so they sought to account for difference as irrationality or immaturity. But Lovejoy is clear about the definition; it is a question of rational property in children.

Nothing a female could do could lead the species across the hominoid–hominid boundary; she was already doing the best nature allowed. "She would have to devote more energy to parenting. But natural selection has already perfected her maternal skills over the millions of years her ancestors have occupied West Africa. There is, however, an untapped pool of reproductive energy in most primate species—the male" (Lovejoy, 1984, p. 26). Through provisioning his now pair-bonded and sedentary mate at a home base with the fruits of plant and small animal gathering, a male could lead the species across the boundary to the origin of man in the assurance of fatherhood. Lovejoy gave up hunting to mark manhood, but he could not dispense with paternity. Mothers could have lots of babies, the role Theodore Roosevelt so hoped for in his 1905 analysis of modern (white) "race suicide," that concept for dawning consciousness of the politics of differential reproduction. The species had reason to stand upright at last, even if not too efficiently at the start. Man was on the long lonesome road. And women's place in this revolution is where it was imagined cross-racially in a fair section of U.S. 1960s politics—prone. As Lovejoy put it, women did not "lose" estrus; they constantly display its signs. For the new strategy to succeed, "the female must remain constantly attractive to the male. . . . While the mystery of bipedality has not been completely solved, the motive is becoming apparent" (Lovejoy, 1984, p. 28). Small wonder that Lovejoy cites his brother-colleague for evidence that human "females are continually sexually receptive" (Lovejoy, 1981, p. 346; footnoted on p. 350: "D.C. Johanson, personal communication").

Why did serious scientists need to respond to this story? Hrdy and Zihlman were involved with their own research and publication, attempting to establish the authority of stories quite different from Lovejoy's, some of them involving Lucy. It took time to write about Lovejoy, just as it took space in this essay, and Lovejoy has not taken the time to write in detail about the interpretations of Hrdy or Zihlman. His decision not to cite Zihlman's substantial and directly pertinent technical analyses, in a paper replete with references, effectively obscured from the readers of the 1981 *Science* cover story her significant work on bipedalism, sexual dimorphism, and reconstructions of hominid social and reproductive behavior at the crucial boundary (Lovejoy, 1981; for summary and previous references, see Zihlman, 1983; Laporte & Zihlman, 1983). The cover article is the point: Lovejoy's story and his involvement with immensely important fossils cannot be ignored. Milton's daughters do not have that luxury. But they do have a weapon more potent than the undecidably lost or omnipresent signs of estrus; they type.

Zihlman responded with Jerrold Lowenstein in a parodic, serious interview with a freeze-thawed, living Australopithecus female fossil: "A Few Words with Ruby" (Zihlman & Lowenstein, 1983). Ruby got her name from the Ruby Tuesday of the Rolling Stones. Interviewed in the British Museum, she discussed the social–reproductive lives of her group, as well as her relationship with her discoverers' scientific friend, Dr. Aaron Killjoy. "Ruby sighed, 'One thing hasn't changed in three million years. Males still think sex explains everything'" (Zihlman & Lowenstein, 1983, p. 83). Ruby was slated for a busy schedule under the patronage of science, including a BBC documentary called "Ruby, Woman of the Pliocene." But she took time to describe her life in terms reminiscent of a contemporary species, *Pan paniscus*, the pygmy chimpanzee, Zihlman's favored model species for studying origins. The essentials of Ruby's account include active, mobile hominid females, even when carrying babies, food-sharing patterns emerging from matrifocal social organization and selection for more sociable males within that context, and open and flexible social groups. Food played a larger role than sex.

Aside from the specifics, there is a formal difference in the Zihlman story, both in the interview with Ruby and elsewhere (Laporte & Zihlman, 1983; Zihlman, 1983). There is no origin of the family. There is no chasm, no expulsion from the garden, no dramatic boundary crossing. The Miocene/Pliocene boundary is depicted as less hostile, more as an opening of possibility for which *paniscus*-like hominoids were ready, socially and physically. There is no narrative of a time of innocence in a forest, followed by a time of trial on the dry plain, calling out the heroics of reproductive politics. The basic narratives of causality depend less on the antagonistic dialectic of nature–culture, the dramatic stories of the West and its others. In the Western sense, there is simply less drama. Zihlman's stories regularly do not generate "others" as raw material for crucial transitions to higher stages. This is not a result of "moral superiority" or "genius"; it is an historical possibility made available by political–scientific struggle to generate coherent accounts of connection. One object of knowledge that falls away in these accounts is "the family." In a sense, there is nothing to explain, no primal scene, whose tragic consequences escalate into history, no civilization and its discontents, no cascading repressions. No wonder the reproductive politics look different.

These basic narrative strategies constrain Zihlman's accounts of both the physical and social parameters of human evolution. They are iconically represented in her *Human Evolution Coloring Book* illustration of Lucy and her relatives, the pygmy chimp and "man." (Zihlman, 1982, p. iv). "Man" here is a female, a visual jolt in the illustration, even allowing Lucy's probable sex. The outline of a tall human figure contains a twinned ape, one half of whom is a pygmy chimp female, the other half, joined at the midline, is a reconstruction of Lucy. The three figures share several body boundaries, while differentiated in degrees of bipedal specialization and other particulars that mark the boundary between hominoid and hominid. There is a play of similarity and difference among the two genera of hominids, *Homo* and *Australopithecus*, and the chimpanzee species, *paniscus*. They model each other in an invitation to the student to color their common space. Boundaries exist in Zihlman's accounts, but they suggest zones of transition rather than the inversions of dualist stories.

In collaboration with a science writer, Hrdy responded to Lovejoy with a popular piece called "What Did Lucy's Husband Stand For?" (Hrdy & Bennett, 1981). This piece and her review of Donald Symons' sociobiological *The Evolution of Human Sexuality* contain the kernel of Hrdy's explanatory strategy and her view of the centrality of reproductive politics in the human place in nature (Hrdy, 1979). Like Zihlman, Hrdy must reconstruct what requires explanation, and the chief casualty is "the family." Also, like Zihlman, her parody of Lovejoy's expulsion from the Garden account, complete with Eve fated to ever more efficient production of babies in conditions of scarcity, is coupled with detailed renegotiation of narratives of sexual politics. But unlike Zihlman, Hrdy sees the nub of what it means to be human in the question of sex. Lovejoy's monogamous nuclear family with male provider and faithful-but-ever-attractive female baby machine is dispatched by comparison of human patterns of sexual dimorphism with other primate patterns in relation to breeding systems and ecological niche; by discussion of female subsistence activity among human gatherer-hunters; by considering ecological explanations based on female feeding possibilities and needs; and by rational genetic investment assessments for a male contemplating monogamy among early hominids. It looks like a bad bet. Hrdy also dispatches the problem of bipedalism, clearing the ground for the important question of active female sexuality. She approves Peter Rodman's explanation that hominoid ancestors were not very efficient quadrupeds, so a transition to inefficient bipedalism was not much of a loss. The most convincing explanations seem to work by unraveling the object of attention.

The politics of the female orgasm is what requires elucidation in Hrdy's narrative logic, not from some errant prurient interest or special pleading for females, but in the interests of rational mind and equal potential for citizenship in the late capitalist primate polis ruled by the logic of the market (Hrdy, 1979). For Symons, the human female orgasm is a by-product of the more perfect and sensible male version so essential to the tale of reproductive maximization strategies in the face of limited resources called *females*. But Hrdy argues that women evolved too; that is, there is large variation in female fitness and so grounds for selection. Female reproductive fitness can vary in at least five categories: female choice of mate, female elucidation of male support or protection, competition with other females for resources, cooperation with other females, and female ergonomic efficiency. The politics of the female orgasm is in the center of the matter. It is part of an active, investing, and calculating female sexuality, where sex is of the essence of mind. Those two categories collapse in sociobiological accounts. Hrdy sees active female sexuality as a tool to manipulate and deceive males, not to enhance a chimeric marriage bond, but to induce male aid, willing or not, in her reproductive process. Concealed ovulation, orgasm, active solicitation when conception is impossible: all these are rational behaviors of an investor in certain market conditions that pertained at the time of origins.

Property in the self has been the ground of citizenship in the West, at least since the 17th century. The tenuousness of such a form of property for reproducing females has made citizenship anomalous or simply impossible for real women. Abortion and other reproductive rights politics today should lay to rest any complacency that the issues are past. Hrdy is arguing a biological form of the struggle for rational citizenship within the constraints of the narrative logic of scarcity and agonistic difference; that is, within the traditional bounds of Western stories. Nineteenth century feminists appropriated the available medical doctrines of the female animal, the creature organized around the uterus, the scene of fruitful production, of nurturing, and argued the rationality of female citizenship in the form of social motherhood, extending the uterine power of the hearth to the sterile masculine public world. Sociobiological feminists carry out a parallel task with 20th century coded bodies and their investment portfolios. Female primates got orgasms in the 1970s because they needed them for a larger political struggle. The active pursuit of pleasure and profit is the mark of rational man, the practice of civic virtue in the state of nature. Woman could do no less. Female sex took on the promising dual property, simultaneously natural and active, that is so potent in Western stories.

So primatology is politics by other means. In myriad mundane ways, primatology is a practice for the negotiation of the possibility of community, of a public world, of rational action. It is the negotiation of the time of origins, the origin of the family, the boundary between self and other, hominid and hominoid, human and animal. Primatology is about the principle of action, mutability, change, energy, about the possibility and constraints of politics. The reading of Lucy's bones is about all those things. In other times and places, people might have cast Lucy's bones in the rituals of necromancy for purposes Western observers called *magical*. But Western people cast her bones into "scientific" patterns for insight into a human future made problematic by the very material working-out of the Western stories of apocalypse and transcendence. The past, the animal, the female, nature: these are the contested zones in the allochronic discourse of primatology.

Acknowledgements—Support for this paper was provided by an Academic Senate Faculty Research Grant from the University of California at Santa Cruz. Thanks especially to the primatologists who allowed me to interview them. In this essay, I am indebted to Jeanne Altmann, Stuart Altmann, Naomi Bishop, Dorothy Cheney, Suzanne Chevalier-Skolnikoff, Irven DeVore, Phyllis Dolhinow, Robert Hinde, Sarah Blaffer Hrdy, Alison Jolly, Peter Marler, Nancy Nicolson, Suzanne Ripley, Thelma Rowell, Robert Seyfarth, Joan Silk, Barbara Smuts, Thomas Struhsaker, Shirley Strum, Jane Teas, Sherwood Washburn, Patricia Whitten, Richard Wrangham, and Adrienne Zihlman.

REFERENCES

Altmann, J. (1974). Observational study of behavior: Sampling methods. *Behaviour, 49*, 227–267.

Altmann, J. (1980). *Baboon mothers and infants*. Cambridge: Harvard University Press.

Asquith, P. (1984). Bases for differences in Japanese and Western primatology. Paper delivered at the 12th Meeting of CAPA/AAPC, University of Alberta, Edmonton, Alberta.

Baranga, D. (1978). The role of nutritive value in the food preferences of the red Colobus and black-and-white Colobus in the Kibale Forest, Uganda. Unpublished master's thesis. Makerere University, Uganda.

Baudrillard, J. (1983). *Simulations*. Translated by Paul Foss, Paul Patton, & Phillip Beitchman. New York: Semiotext(e).

Beer, G. (1983). *Darwin's plots*. London: Routledge and Kegan Paul.

Burton, F. (1971). Sexual climax in female *Macaca mulatta*. *Proceedings of the Third International Congress of Primatology, 3*, 180–191.

Callon, M., & Latour, B. (1981). Unscrewing the big leviathan or how do actors microstructure reality? In K. Knorr-Cetina & A. Cicourel (Eds.), *Advances in social theory and methodology: Toward an integration of micro and macro sociologies*. London: Routledge and Kegan Paul.

Cheney, D., Seyfarth, R., Smuts, B., Struhsaker, T., & Wrangham, R. (Eds.). (forthcoming) *Primate Societies*.

Chevalier-Skolnikoff, S. (1974). Male-female, female-female, and male-male sexual behavior in the stumptail monkey, with special attention to the female orgasm. *Archives of Sexual Behavior, 3*, no. 2, 95–116.

Chodorow, N. (1978). *The reproduction of mothering*. Los Angeles: University of California Press.

Clifford, J. (1983). On ethnographic authority. *Representations, 1*, no. 2, 118–146.
Dawkins, R. (1976). *The selfish gene*. London: Oxford University Press.
Delaney, C. (1985). Virgin birth, once again. Unpublished manuscript.
DeVore, I. (Ed.). (1965). *Primate behavior: Field studies of monkeys and apes*. New York: Holt, Rinehart and Winston.
de Waal, F. (1982). *Chimpanzee politics: Power and sex among the apes*. New York: Harper and Row.
Douglas, M. (1970). *Natural symbols*. London: Cresset.
Elgin, S. H. (1984). *Native Tongue*. New York: Daw.
Fabian, J. (1983). *Time and the other*. New York: Columbia University Press.
Fedigan, L. M. (1982). *Primate paradigms: Sex roles and social bonds*. Montreal: Eden Press.
Fossey, D. (1983). *Gorillas in the mist*. Boston: Houghton Mifflin.
Foucault, M. (1978). *The history of sexuality: Introduction*. New York: Pantheon.
Goodall, J. van L. (1971). *In the shadow of man*. Boston: Houghton Mifflin.
Goodall, J., Bandoro, A., Bergmann, E., Busse, C., Matama, H., Mpongo, E., Pierce, A., & Riss, D. (1979). Intercommunity interactions in the chimpanzee population of the Gombe National Park. In D. A. Hamburg, & E. R. McCown (Eds.), *The great apes*, pp. 13–54. Menlo Park, CA: Benjamin/Cummings.
Gubar, S., & Gilbert, S. (1979). *Madwoman in the attic*. New Haven, CT: Yale University Press.
Haraway, D. (1979). The biological enterprise: Sex, mind, and profit from human engineering to sociobiology. *Radical History Review*, no. 20, 206–237.
Haraway, D. (1981). In the beginning was the word: The genesis of biological theory. *Signs, 6*, no. 3, 469–481.
Haraway, D. (1981–1982). The High Cost of Information in Post World War II Evolutionary Biology. *Philosophical Forum, XIII*, nos. 2–3, 244–278.
Haraway, D. (1983a). Signs of dominance: From a physiology to a cybernetics of primate society. *Studies in History of Biology, 6*, 129–219.
Haraway, D. (1983b). The contest for primate nature: Daughters of man the hunter in the field, 1960–1980. In M. Kann (Ed.), *The future of American democracy*, pp. 175–207. Philadelphia, PA: Temple University Press.
Haraway, D. (1985). A manifesto for Cyborgs: Science, technology, and socialist feminism in the 1980s. *Socialist Review*, no. 80, 65–108.
Harding, S. (1983). Why has the sex/gender system become visible only now? In S. Harding, & M. Hintikka (Eds.), *Discovering reality: Feminist perspectives on epistemology, metaphysics, methodology, and philosophy of science*, pp. 311–324. Dordrecht, The Netherlands: Reidel.
Harding, S. (1985). *The science question in feminism*. Ithaca, NY: Cornell University Press.
Harris, N. (1973). *Humbug: The art of P.T. Barnum*. Boston: Little, Brown, & Co.
Hartsock, N. (1983a). *Money, sex and power*. New York: Longman.
Hartsock, N. (1983b). The feminist standpoint: Developing the ground for a specifically feminist historical materialism. In S. Harding & M. Hintikka (Eds.), *Discovering reality: Feminist perspectives on epistemology, metaphysics, methodology, and philosophy of science*, pp. 283–210. Dordrecht, The Netherlands: Reidel.
Hausfater, G., & Hrdy, S. B. (Eds.). (1984). *Infanticide: Comparative and evolutionary perspectives*. Chicago: Aldine.
Hinde, R. (Ed.). (1983). *Primate social relationships*. Sunderland, MA: Sinauer.
Hrdy, S. B. (1979). The evolution of human sexuality: The latest word and the last. *The Quarterly Review of Biology, 54*, 309–314.

Hrdy, S. B. (1981). *The woman that never evolved*. Cambridge: Harvard University Press.

Hrdy, S. B. (1983, October). Heat lost. *Science 83*, pp. 73–78.

Hrdy, S., & Bennett, W. (1981, July-August). Lucy's husband: What did he stand for? *Harvard Magazine*, pp. 7–9, 46.

Jameson, F. (1981). *The political unconscious: Narrative as a socially symbolic act*. Ithaca, NY: Cornell University Press.

Jameson, F. (1984, July/August). Post modernism or the cultural logic of late capitalism. *New Left Review*, pp. 53–94.

Jay, P. (Ed.). (1968). *Primates: Studies in adaptation and variability*. New York: Holt, Rinehart and Winston.

Johanson, D., & Edey, M. (1981). *Lucy: The beginnings of humankind*. New York: Simon and Schuster.

Keller, E. F. (1983). *A feeling for the organism*. New York: W. H. Freeman.

Keller, E. F. (1985). *Reflections on gender and science*. New Haven, CT: Yale University Press.

Lancaster, J. (1979). Sex and gender in evolutionary perspective. In H. A. Katchadourian (Ed.), *Human sexuality: A comparative and developmental approach*. Los Angeles: University of California Press.

Landau, M. (1984). Human evolution as narrative. *American Scientist, 72*, 262–268.

Laporte, L. F., & Zihlman, A. L. (1983). Plates, climate and hominoid evolution. *South African Journal of Science, 79*, 96–109.

Latour, B. (1978). Observing scientists observing baboons observing. . . . Paper prepared for the Wenner Gren conference, "Baboon Field Research: Myths and Models." Unpublished manuscript.

Latour, B. (1984). *Les microbes, guerre et paix, suivi de irreductions*. Paris: Metaillie.

Latour, B., and Strum, S. (1983). Oh, please, tell us another story. Unpublished manuscript.

Lovejoy, C. O. (1981). The origin of man. *Science, 211*, no. 4480, 341–350.

Lovejoy, C. O. (1984, October). The natural detective. *Natural History*, pp. 24–28.

Morris, D. (1967). *The naked ape*. New York: McGraw-Hill.

Nash, R. (1982). *Wilderness and the American mind*. (3rd ed.). New Haven, CT: Yale University Press.

Ortner, S. B. (1972). Is female to male as nature is to culture? *Feminist Studies, 1*, 5–31. Reprinted in M. Z. Rosaldo & L. Lamphere (Eds.), *Woman, culture, and society*. pp. 67–88. Stanford, CT: Stanford University Press.

Rosaldo, M. Z. (1980). The use and abuse of anthropology. *Signs, 5*, no. 3.

Rowell, T. (1974). The concept of dominance. *Behavioral Biology, 11*, 131–154.

Ruddick, S. (1982). Maternal Thinking. In B. Thorne & M. Yalom (Eds.), *Rethinking the family*, pp. 76–94. New York: Longmann.

Said, E. (1978). *Orientalism*. New York: Pantheon.

Sandoval, C. (n.d.). *Women respond to racism*. Oakland, CA: Center for Third World Organizing.

Sandoval, C. (1984). Disillusionment and the poetry of the future: The making of oppositional consciousness. Unpublished doctoral qualifying essay. History of Consciousness Department, University of California at Santa Cruz.

Sherfey, M. J. (1973). *The nature and evolution of female sexuality*. New York: Vintage.

Shostak, M. (1981). *Nisa: The life and words of a !Kuno woman*. Cambridge: Harvard University Press.

Small, M. (Ed.). (1984). *Female primates: Studies by women primatologists*. New York: Alan Liss.
Strathern, M. (1980). No nature, no culture: The Hagan case. In M. Strathern & C. MacCormmach (Eds.), *Nature, Culture, Gender*. London: Oxford University Press.
Strathern, M. (1984). Dislodging a world view: Challenge and counter-challenge in the relationship between feminism and anthropology. Unpublished manuscript from lecture in the series, *Changing paradigms: The impact of feminist theory on the world of scholarship*. Research Centre for Women's Studies, Adelaide, Australia.
Symons, D. (1979). *The evolution of human sexuality*. New York: Oxford University Press.
Tiger, L., & Fox, R. (1971). *The imperial animal*. New York: Holt, Rinehart and Winston.
Tournier, M. (1972). *Vendredi ou les limbes du Pacifique*. Paris: Gallimard.
Traweek, S. (1982). *Uptime, downtime, spacetime, and power: An ethnographic study of the high energy physics community in Japan and the United States*. Unpublished doctoral dissertation. History of Consciousness Board, University of California at Santa Cruz.
Trescott, M. (1984). Women engineers in history: Profiles in holism and persistence. In V. Haas & C. Perrucci (Eds.), *Women in Scientific and Engineering Professions*, pp. 181–204. Ann Arbor: University of Michigan Press.
Trivers, R. (1972). Parental investment and sexual selection. In B. Campbell (Ed.), *Sexual selection and the descent of man*, pp. 136–179. Chicago: Aldine.
Wasser, S. (Ed.). (1983). *Social behavior of female vertebrates*. New York: Academic Press.
Williams, G. (1966). *Adaptation and natural selection*. Princeton, NJ: Princeton University Press.
Wilson, E. O. (1975). *Sociobiology, the new synthesis*. Cambridge, MA: The Belknap Press of Harvard University Press.
Winner, L. (1980, Winter). Do artifacts have politics? *Daedalus*, pp. 121–136.
Winner, L. (1983). Techne and politeia: The technical constitution of society. In T. Durbin & F. Rapp (Eds.), *Philosophy and technology*, pp. 97–111. Dordrecht, The Netherlands: Reidel.
Wrangham, R. (1979). On the evolution of ape social systems. *Social Science Information, 18*, no. 3, 335-369.
Zacharias, K. (1984). The Owen-Huxley debate on the brain: A new appraisal. Unpublished manuscript.
Zihlman, A. (1982). *The human evolution coloring book*. New York: Barnes and Noble.
Zihlman, A. (1983). A behavioral reconstruction of Australopithecus. In K. J. Reichs (Ed.), *Hominid origins: Inquiries past and present*, pp. 207–238. Washington, DC: University Press of America.
Zihlman, A., & Lowenstein, J. (1983, April 14). A few words with Ruby. *New Scientist*, pp. 81–83.
Zuckerman, S. (1932). *The social life of primates*. London: Routledge.

Chapter 6

Empathy, Polyandry, and the Myth of the Coy Female

Sarah Blaffer Hrdy

> Sexual selection theory (Bateman, 1948; Darwin, 1871; Trivers, 1972; Williams, 1966) is one of the crown jewels of the Darwinian approach basic to sociobiology. Yet so scintillating were some of the revelations offered by the theory, that they tended to outshine the rest of the wreath and to impede comprehension of the total design, in this instance, the intertwined, sometimes opposing, strategies and counter strategies of both sexes which together compose the social and reproductive behavior of the species. (Hrdy & Williams, 1983, p. 7)

But why did that happen, and how? And what processes led to the current destabilization of the model and reformulation of our thinking about sexual selection?

INTRODUCTION

For over three decades, a handful of partially true assumptions were permitted to shape the construction of general evolutionary theories about sexual selection. These theories of sexual selection presupposed the existence of a highly discriminating, sexually "coy," female who was courted by sexually undiscriminating males. Assumptions underlying these stereotypes included, first, the idea that relative male contribution to offspring was small, second, that little variance exists in female reproductive success compared to the very great variance among males, and third, that fertilization was the only reason for females to mate. While appropriate in some contexts, these conditions are far from universal. Uncritical acceptance of such assumptions has greatly hampered our understanding of animal breeding systems particularly, perhaps, those of primates.

These assumptions have only begun to be revised in the last decade, as researchers began to consider the way Darwinian selection operates on females as well as males. This paper traces the shift away from the stereotype of female as sexually passive and discriminating to current models in which females are seen to play an active role in managing sexual consortships that go beyond traditional "mate choice." It is impossible to understand this history without taking into account the background, including the gender, of the researchers involved. Serious consideration is given to the possibility that the empathy for other females subjectively felt by women researchers may have been instrumental in expanding the scope of sexual selection theory.

ANISOGAMY AND THE BATEMAN PARADIGM

In one of the more curious inconsistencies in modern evolutionary biology, a theoretical formulation about the basic nature of males and females has persisted for over three decades, from 1948 until recently, despite the accumulation of abundant openly available evidence contradicting it. This is the presumption basic to many contemporary versions of sexual selection theory that males are ardent and sexually undiscriminating while females are sexually restrained and reluctant to mate. My aims in this paper will be to examine this stereotype of "the coy female," to trace its route of entry into modern evolutionary thinking and to examine some of the processes that are only now, in the last decade, causing us to rethink this erroneous corollary to a body of theory (Darwin, 1871) that has otherwise been widely substantiated. In the course of this examination, I will speculate about the role that empathy and identification by researchers with same-sex individuals may have played in this strange saga.

Obviously, the initial dichotomy between actively courting, promiscuous males and passively choosing, monandrous females dates back to Victorian times. "The males are almost always the wooers," Darwin wrote in 1871, and he was very clear in his own writings that the main activity of females was to choose the single best suitor from among these wooers. As he wrote in *The Descent of Man and Selection in Relation to Sex* (1871), "It is shown by various facts, given hereafter, and by the results fairly attributable to sexual selection, that the female, though comparatively passive, generally exerts some choice and accepts one male in preference to the others." However the particular form in which these ideas were incorporated into modern and ostensibly more "empirical" versions of post-Darwinian evolutionary thought derived from a 1948 paper about animals by a distinguished plant geneticist, Angus John Bateman.

Like so much in genetics, Bateman's ideas about the workings of nature were based primarily on experiments with *Drosophila*, the minuscule flies that materialize in the vicinity of rotting fruit. Among the merits of fruitflies rarely appreciated by housekeepers are the myriad of small genetic differences that determine a fruitfly's looks. Bred over generations in a laboratory, distinctive strains of *Drosophila* sporting odd-colored eyes, various bristles, peculiar crenulations here and there, grotesquely shaped eyes, and so forth can be produced by scientists, and these markers are put to use in tracing genealogies.

Bateman obtained various lots of differently decorated *Drosophila* all belonging to the one species, *Drosophila melanogaster*. He housed three to five flies of each sex in glass containers and allowed them to breed. On the basis of 64 such experiments, he found (by counting the offspring bearing their parents' peculiar genetic trademarks) that while 21% of his males failed to fertilize any female, only 4% of his females failed to produce offspring.

A highly successful male, he found, could produce nearly three times as many offspring as the most successful female. Furthermore, the difference between the most successful and the least successful male, what is called the *variance* in male reproductive success, was always far greater than the variance among females. Building upon these findings, Bateman constructed the centerpiece to his paradigm: whereas a male could always gain by mating just one more time, and hence benefit from a nature that made him undiscriminatingly eager to mate, a female, already breeding near capacity after just one copulation, could gain little from multiple mating and should be quite uninterested in mating more than once or twice.

From these 64 experiments with *Drosophila*, Bateman extrapolated to nature at large: selection pressures brought about by competition among same-sexed individuals for representation in the gene pools of succeeding generations would almost always operate more strongly upon the male than upon the female. This asymmetry in breeding potential would lead to a nearly universal dichotomy in the sexual nature of the male and female:

> One would therefore expect to find in all but a few very primitive organisms . . . that males would show greater intra-sexual selection than females. This would explain why . . . there is nearly always a combination of an undiscriminating eagerness in the males and a discriminating passivity in the females. Even in a derived monogamous species (e.g. man) this sex difference might be expected to persist as a rule. (Bateman, 1948, p. 365)

This dichotomy was uncritically incorporated into modern thinking about sexual selection. In his classic 1972 essay on "Parental Investment and Sexual Selection," Harvard biologist Robert Trivers acknowledged Bateman's paper as "the key reference" (provided him, as it happens by one of the major evolutionary biologists of our time, and Trivers' main mentor at Harvard, Ernst Mayr). Trivers' essay on parental investment, carrying with it Bateman's model, was to become the second most widely cited paper in all of sociobiology, after Hamilton's 1964 paper on kin selection.

Expanding on Bateman's original formulation, Trivers argued that whichever sex invests least in offspring will compete to mate with the sex investing most. At the root of this generalization concerning the sexually discriminating female (apart from Victorian ideology at large) is the fact of anisogamy (gametes unequal in size) and the perceived need for a female to protect her already substantial investment in each maternal gamete; she is under selective pressure to select the best available male to fertilize it. The male, by contrast, produces myriad gametes (sperm), which are assumed to be physiologically cheap to produce (note, however, that costs to males of competing for females are rarely factored in), and he disseminates them indiscriminately.

Two central themes in contemporary sociobiology then derive directly from Bateman. The first theme is the dichotomy between the "nurturing female," who invests very much more per offspring than males, and "the competitive male," who invests little or nothing beyond sperm but who actively competes for access to any additional female (see for example Daly & Margo Wilson,* 1983, pp. 78–79; Trivers, 1985, p. 207). As Trivers noted in his summary of Bateman's experiments with *Drosophila*, "A female's reproductive success did not increase much, if any, after the first copulation and not at all after the second; most females were uninterested in copulating more than once or twice" (1972, p. 138). And so it was that "coyness" came to be the single most commonly mentioned attribute of females in the literature on sociobiology. Unlike the male, who, if he makes a mistake can move on to another female, the female's investment was initially considered to be so great that she was constrained from aborting a bad bet and attempting to conceive again. (Criticisms and recent revisions of the notion are discussed later in the section, "The Females Who Forgot to be Coy".) In this respect, contemporary theory remains fairly faithful to Darwin's original (1871) two-part definition of sexual selection. The first part of the theory predicts competition between males for mates; the second, female choice of the best competitor.

The second main sociobiological theme to derive from Bateman is not explicitly discussed in Darwin but is certainly implicit in much that Darwin wrote (or more precisely, did not write) about females. This is the notion that female investment is already so large that it can not be increased and the idea that most females are already breeding close to capacity. If this were so, the variance in female reproductive success would be small, making one female virtually interchangeable with another. A logical corollary of this notion is the incorrect conclusion that selection operates primarily on males.

The conviction that intrasexual selection will weigh heavily upon males while scarcely affecting females was explicitly stated by Bateman, but also appears in implicit form in the writings of contemporary sociobiologists (Daly & Margo Wilson, 1983, Chapter 5; Wilson, 1978, p. 125). It is undeniable that males have the capacity to inseminate multiple females while females (except in species such as those squirrels, fish, insects, and cats, where several fathers can sire a single brood) are inseminated – at most – once each breeding

*In this chapter, I designate women researchers by spelling out their first names; the point of using this admittedly odd convention will become clear in the section on "The Role of Women Researchers."

period. But a difficulty arises when the occasionally true assumption that females are not competing among themselves to get fertilized is then interpreted to mean that there will be reduced within-sex competition among females generally (e.g., Freedman, 1979, p. 33).

Until about 1980—and even occasionally after that—some theoreticians were writing about females as though each one was relatively identical in both her reproductive potential and in her realization of that potential. This erroneous generalization lead some workers (perhaps especially those whose training was not in evolutionary biology per se) to the erroneous and patently non-Darwinian conclusion that females are not subject to selection pressure at all and the idea that competition among males is somehow more critical because "leaving offspring is at stake" (Carol Cronin, 1980, p. 302; see also Virginia Abernethy, 1978, p. 132). To make an unfortunate situation worse, the close conformity between these notions and post-Victorian popular prejudice meant that ideas about competitive, promiscuous men and choosey women were selectively picked up in popular writing about sociobiology. An article in *Playboy Magazine* celebrating "Darwin and the Double Standard" (Morris, 1979) comes most vividly to mind, but there were many others.

THE FEMALES WHO FORGOT TO BE COY

Field studies of a number of animal groups provide abundant examples of females who, unlike Bateman's *Drosophila*, ardently seek to mate more than once or twice. Furthermore, fertilization by the best male can scarcely be viewed as their universal goal since in many of these cases females were not ovulating or else were actually pregnant at the time they solicit males.

It has been known for years (among some circles) that female birds were less than chaste, especially since 1975 when Bray, Kennelly, and Guarino demonstrated that when the "master" of the blackbird harem was vasectomized, his females nevertheless conceived (see also Lumpkin, 1983). Evelyn Shaw and Joan Darling (1985) review some of this literature on "promiscuous" females, particularly for marine organisms. Among shiner perch, for example, a female who is not currently producing eggs will nevertheless court and mate with numbers of males, collecting from each male sperm that are then stored in the female's ovaries till seasonal conditions promote ovulation. Female cats, including leopards, lions, and pumas are notorious for their frequency of matings. A lioness may mate 100 times a day with multiple partners over a 6–7-day period each time she is in estrus (Eaton, 1976). Best known of all, perhaps, are such primate examples as savanna baboons, where females initiate multiple brief consortships, or chimpanzees, where females alternate between prolonged consortships with one male and communal mating with all males in the vicinity (DeVore, 1965; Hausfater, 1975; Caroline Tutin, 1975). However, only since 1979 or so has female promiscuity been a subject of much theoretical interest (see for example Alatalo, Lundberg, & Stahlbrandt, 1982; Sandy Andelman, in press; Gladstone, 1979; Sarah Blaffer Hrdy, 1979; Susan Lumpkin, 1983; Meredith Small, forthcoming; R. Smith, 1984; Wirtz, 1983), largely I believe because theoretically the phenomenon should not have existed and therefore there was little theoretical infrastructure for studying it, certainly not the sort of study that could lead to a PhD (or a job).

In terms of the order Primates, evidence has been building since the 1960s that females in a variety of prosimian, monkey, and ape species were managing their own reproductive careers so as actively to solicit and mate with a number of different males, both males within their (supposed) breeding unit and those outside it. As theoretical interest increased, so has the quality of the data.

But before turning to such evidence, it is first critical to put sex in perspective. To correct the stereotype of "coyness," I emphasize female sexual activity but, as always in such debates, reality exists in a plane distinct from that predefined by the debate. In this case, reality is hours and hours, sometimes months and months, of existence where sexual behavior is not even an issue, hours where animals are walking, feeding, resting, grooming. Among baboons (as in some human societies) months pass when a pregnant or lactating mother engages in no sexual behavior at all. The same is generally true for langurs, except that females under particular conditions possess a *capacity* to solicit and copulate with males even if pregnant or lactating, and they sometimes do so. At such times, the patterning of sexual receptivity among langurs could not be easily distinguished from that of a modern woman. The same could be said for the relatively noncyclical, semicontinuous, situation-dependent receptivity of a marmoset or tamarin.

With this qualification in mind—that is the low frequencies of sexual behavior in the lives of *all* mammals, who for the most part are doing other things—let's consider the tamarins.

Tamarins are tiny South American monkeys, long thought to be monogamous. Indeed, in captivity, tamarins do breed best when a single female is paired with one mate. Add a second female and the presence of the dominant female suppresses ovulation in the subordinate. (The consequences of adding a second male to the cage are unknown, since such an addition was thought to violate good management practices.) Nevertheless, in the recent (and first) long-term study of individually marked tamarins in the wild, Anne Wilson Goldizen discovered that given the option, supposedly monogamous saddle-backed tamarins (*Saguinus fusicollis*) will mate with several adult males, each of whom subsequently help to care for her twin offspring in an arrangement more nearly "polyandrous" than monogamous (Goldizen & Terborgh, forthcoming). Furthermore the presence of additional males, and their assistance in rearing young may be critical for offspring survival.) One of the ironies here, pointed out in another context by Janet Sayers (1982), is that females are thus presumed to commit what is known in sociobiology as a *Concorde fallacy*; that is, pouring good money after bad. Although in other contexts (e.g., Dawkins, 1976) it has been argued that creatures are selected to cut bait rather than commit Concorde fallacies, mothers were somehow excluded from this reasoning (however, see Trivers, 1985, p. 268, for a specific acknowledgement and correction of the error). I happen to believe that the resolution to this contradiction lies in recognizing that gamete producers and mothers do indeed "cut bait" far more often than is generally realized, and that skipped ovulations, spontaneous abortion, and abandonment of young by mothers are fairly routine events in nature. That is, the reasoning about the Concorde fallacy is right enough, but our thinking about the commitment of mothers to nurture no matter what has been faulty.)

Indeed, on the basis of what I believe today (cf. Hrdy, 1981, p. 59), I would argue that a polyandrous component* is at the core of the breeding systems of most troop-dwelling primates: females mate with many males, each of whom may contribute a little bit toward the survival of offspring. Barbary macaques provide the most extreme example (Taub, 1980), but the very well-studied savanna baboons also yield a similar, if more moderate, pattern. David Stein (1981) and Jeanne Altmann (1980) studied the complex interactions between adult males and infants. They found that (as suggested years ago by Tim Ransom and Bonnie Ransom, 1972) former, or sometimes future, consorts of the mother develop special relationships with that female's infant, carrying it in times of danger and protecting it from conspecifics, possibly creating enhanced feeding opportunities for the infant. These relationships are made possible by the mother's frequent proximity to males with whom she has special relationships and by the fact that the infant itself comes to trust these males and seek them out; more is at issue than simply male predilections. Altmann aptly refers to such males as *god-fathers*. Infants, then, are often the focal-point of elaborate male–female–infant relationships, relationships that are often initiated by the females themselves (Barbara Smuts, 1985).

*For want of a better term, *polyandrous* is used here to refer to a female with more than one established mate. The term *promiscuous* will be used to refer to multiple, brief consortships, some of which may last longer. Whereas *polyandrous* is a poor term because it suggests some stable, institutionalized relationship, which is probably wrong for describing tamarins, *promiscuous* is also problematic. It implies a lack of selectivity among females, which may or may not be the case. Davies and Lundberg (1984, p. 898) have recently proposed using the term *polygnandry* to refer to "two or three males sharing access to two, three or four females." Such a term applies to Barbary macaques and might be a good one for the baboon situation except that there is not a 100% overlap in the females with which each male mates. Clearly, the terminology needs to be cleared up, but for the time being the important point is to emphasize the contrast between what we now know and the old stereotype of monandrous females selecting a single mate.

Even species such as Hanuman langurs, blue monkeys, or redtail monkeys, all primates traditionally thought to have "monandrous" or "uni-male" breeding systems, are far more promiscuous than that designation implies. Indeed, mating with outsiders is so common under certain circumstances as to throw the whole notion of one-male breeding units into question (Cords, 1984; Tsingalia & Thelma Rowell, 1984). My own first glimpse of a langur, the species I was to spend nearly 10 years studying intermittently, was of a female near the Great Indian Desert in Rajasthan moving rapidly through a steep granite canyon, moving away from her natal group to approach and solicit males in an all-male band. At the time, I had no context for interpreting behavior that merely seemed strange and incomprehensible to my Harvard-trained eyes. Only in time, did I come to realize that such wandering and such seemingly "wanton" behavior were recurring events in the lives of langurs.

In at least three different sets of circumstances female langurs solicit males other than their so-called *harem-leaders*: first, when males from nomadic all-male bands temporarily join a breeding troop; second, when *females* leave their natal troops to travel temporarily with all-male bands and mate with males there; and third, when a female for reasons unknown to any one, simply takes a shine to the resident male of a neighboring troop (Hrdy 1977; Moore 1985; filmed in Hrdy, Hrdy, & Bishop, 1977). It may be to abet langurs in such projects that nature has provided them attributes characteristic of relatively few mammals. A female langur exhibits no visible sign when she is in estrus other than to present to a male and to shudder her head. When she encounters strange males, she has the capacity to shift from cyclical receptivity (that is, a bout of heat every 28 days) into a state of semicontinuous receptivity that can last for weeks. Monkeys with similar capacities include vervets, several of the guenons, and gelada baboons, to mention only a few (reviewed in Hrdy & Whitten, 1986).

A number of questions are raised by these examples. First, just exactly why might females bother to be other than coy, that is why should they actively seek out partners including males outside of their apparent breeding units (mate "promiscuously", seek "excess" copulations, beyond what are necessary for fertilization)? Second, why should this vast category of behaviors be, until recently, so generally ignored by evolutionary theorists? As John Maynard Smith noted, in the context of mobbing behavior by birds, "behavior so widespread, so constant, and so apparently dangerous calls for a functional explanation" (1984, p. 294).

To be fair, it should be acknowledged that mobbing behavior in birds is more stereotyped than sexual behavior in wild cats or monkeys, and it can be more systematically studied. Nevertheless, at issue here are behaviors exhibited by the majority of species in the order primates, the best studied order of animals in the world, and the order specifically included by Bateman in his extrapolation from coyness in arthropods to coyness in anthropoids. Furthermore, females engaged in such "promiscuous matings" entail obvious risks ranging from retaliatory attacks by males, venereal disease, the energetic costs of multiple solicitations, predation risks from leaving the troop, all the way to the risk of lost investment by a male consort who has been selected to avoid investing in other males' offspring (Trivers, 1972). In retrospect, one really does have to wonder why it was nearly 1980 before promiscuity among females attracted more than cursory theoretical interest.

Once the initial conceptual block was overcome (and I will argue in the last section that the contributions of women researchers was critical to this phase, at least in primatology), once it was recognized that oh yes, females mate promiscuously and this is a most curious and fascinating phenomenon, the question began to be vigorously pursued. (Note though that the focus of this paper is on male-centered theoretical formulations, readers should be aware that there are other issues here, such as the gap between theoreticians and fieldworkers, which I do not discuss.)

In my opinion, no conscious effort was ever made to leave out female sides to stories. The Bateman paradigm was very useful, indeed theoretically quite powerful, in explaining such phenomena as male promiscuity. But, although the theory was useful in explaining male behavior, by definition (i.e., *sexual selection* refers to competition between one sex for *access* to the other sex) it excluded much within-sex reproductive competition among females, which was not over fertilizations per se but which also did not fall neatly into the realm of the survival-related phenomena normally considered as due to natural selection. (The evolution of sexual swellings might be an example of a phenomenon that fell between definitional cracks and hence went unexplained until recently [Clutton-Brock & Harvey, 1976; Hrdy, 1981].) To understand female promiscuity, for example, we first needed to recognize the limitations of sexual selection theory and then needed to construct a new theoretical base for explaining selection pressures on females.

The realization that male–male competition and female choice explains only a small part of the evolution of breeding systems has led to much new work (e.g., Wasser, 1983, and work reviewed therein). We now have, for example, no fewer than six different models to explain how females might benefit from mating with different males (see Smith, R., 1984, for a recent review).

These hypotheses, most of them published in 1979 or later, can be divided into two categories, first those postulating genetic benefits for the offspring of sexually assertive mothers, and second, those postulating nongenetic benefits for either the female herself or her progeny. All but one of these (the oldest, "prostitution hypothesis") was arrived at by considering the world from a female's point of view.

Whereas all the hypotheses specifying genetic benefits predict that the female should be fertile when she solicits various male partners (except in those species where females have the capacity to store sperm), this condition is not required for the nongenetic hypotheses. It should be noted, too, that only functional explanations for multiple matings are listed. The idea that females simply "enjoy" sex begs the question of why females in a genus such as *Drosophila* do not appear highly motivated to mate repeatedly, while females in other species apparently are so motivated and have evolved specific physiological apparatus making promiscuity more likely (e.g., a clitoris, a capacity for orgasm brought about by prolonged or multiple sources of stimulation, a capacity to expand receptivity beyond the period of ovulation, and so forth; see Hrdy, 1981, Chapter 7 for discussion). Nevertheless, the possibility persists that promiscuous behaviors arise as endocrinological accidents or perhaps that females have orgasms simply because males do (Symons, 1979), and it is worth remembering that an act of faith is involved in assuming that there is any function at all. (I mention this qualifier because I am not interested in arguing a point that can not currently be resolved.)

Assuming that promiscuous behaviors and the physiological paraphernalia leading to them have evolved, four hypotheses are predicated on genetic benefits for the offspring of sexually assertive mothers: (a) the "fertility back-up hypothesis," which assumes that females will need sperm from a number of males to assure conception (Meredith Small, forthcoming; Smith, R., 1984); (b) "the inferior cuckold hypothesis," in which a female paired with an inferior mate surreptitiously solicits genetically superior males when conception is likely (e.g., Benshoof & Thornhill, 1979); (c) "the diverse paternity" hypothesis, whereby females confronted with unpredictable fluctuations in the environment produce clutches sired by multiple partners to diversify paternity of offspring produced over a lifetime (Parker, 1970; Williams, 1975); or (d) in a somewhat obscure twist of the preceding, females in species where litters can have more than one father alter the degree of relatedness between sibs and maternal half-sibs by collecting sperm from several fathers (Davies & Boersma, 1984).

The remaining explanations are predicated on nongenetic benefits for females and do not assume the existence of either genetic differences between males or the existence of female capacities to detect them: (e) the "prostitution" hypothesis, whereby females are thought to exchange sexual access for resources, enhanced status, etc. – the oldest of all the explanations (first proposed by Sir Solly Zuckerman 1932, recently restated by Symons, 1979; see also, Nancy Burley & Symanski, 1981, for discussion); (f) the "therapeutic hypothesis" that multiple matings and resulting orgasm are physiologically beneficial to females or make conception more likely (Mary Jane Sherfey, 1973); (g) the "keep 'em around" hypothesis whereby females (with the connivance of dominant males in the group) solicit subordinate males to discourage these disadvantaged animals from leaving the group (Stacey, 1982); and (h) the "manipulation hypothesis," suggesting that females mate with a number of males in order to confuse information available to males about paternity and thereby extract investment in, or tolerance for, their infants from different males (Hrdy, D. B., 1979; Stacey, 1982).

It is this last hypothesis that I now want to focus on, not because that hypothesis is inherently any better than others, but because I know the most about it and about the assumptions that needed to be changed before it could be dreamed up.

The "manipulation hypothesis," first conceived in relation to monkeys, grew out of a dawning awareness that, first of all, individual females could do a great deal that would affect the survival of their offspring, and second, that males, far from mere dispensers of sperm, were critical features on the landscape where infants died or survived. That is, females were more political, males more nurturing (or at least not neutral), than some earlier versions of sexual selection theory would lead us to suppose.

A FEMALE IS NOT A FEMALE IS NOT A FEMALE

To his credit, A. J. Bateman was a very empirical scientist. He was at pains to measure "actual" and not just "potential" genetic contribution made by parents. Not for him the practice—still prevalent in primatology several decades later—of counting up some male's copulations and calling them *reproductive success*. Bateman counted offspring actually produced. And, in a genus such as *Drosophila*, where infant mortality is probably fairly random and a stretch of bad weather accounts for far more deaths than a spate of bad parenting, the assumption that one mother is equivalent to another mother is probably not farfetched. Such factors as the social status of the mother, her body size, her expertise in child-rearing, or the protection and care elicited from other animals may indeed make little difference. But what if he had been studying monkeys or even somebody's favorite fish? Even for *Drosophila* conditions exist in which females benefit from multiple copulations. In a series of experiments with *Drosophila pseudoobscura*, Turner and Anderson (1983) have shown that the number of offspring that survive to maturity was significantly higher for females allowed to mate for longer periods and with more partners than for females isolated from males after brief mating periods. This effect was most pronounced in laboratory groups that were nutritionally stressed.

The female coho salmon buries her eggs in nests, which she guards for as long as she lives. Females fight over the best nest sites, and about one out of three times, a female will usurp another female's nest and destroy her eggs. Females vary greatly in size, and their differing dimensions may be translated into different degrees of fecundity. A big female may produce more than three times as many eggs as a small one. Differences in the survival of eggs to hatching lead to even greater variance in female reproductive success; there may be as much as a 30-fold difference in number of surviving offspring (Van den Berghe, 1984).

But the mother salmon only breed once; consider an iteroparous monkey mother who, although she produces only one or two infants at a time, breeds over many years and who, like a macaque or baboon, may inherit her feeding range and troop rank from her mother at birth. These legacies will affect her reproductive output and will, in turn, pass to her own daughters. Males of course enter this system, and vary among themselves, but in most instances they are transients, breeding briefly, and indeed, possibly living shorter lives on average than females. Take the extreme example of the gelada baboon who has only one chance for controlling access to a small "harem" of females (who by the way have about as much to do with controlling the male, as he does in controlling them). The male gelada baboon breeds in his unit for several years before another male enters, pushing him into forced retirement. The former "harem-leader" lingers on in the troop, but as a celibate watcher, possibly babysitting, but breeding no more (Dunbar, 1984). It is a tale of the tortoise and the hare. After the male hare is dismissed, the female tortoise breeds on year after year.

Although we do not yet have data on the lifetime reproductive success of males or females from any species of wild primate, I will be surprised if the variance among males exceeds the variance among females by as much as traditionally thought in species such as Japanese or rhesus macaques or gelada baboons. In the most polyandrous species, such as tamarins, variance in the reproductive success of twin-producing females may actually be greater than that for males. If we carry out our calculations over generations, remembering that every male, however wildly reproductively successful, has a mother and a grandmother (e.g., see Hartung, in press) differences in the degree of variance between the two sexes grow even smaller, though extremes of variance in reproductive success will of course crop up one generation sooner for fathers than for mothers.

The anisogamy paradigm of Bateman offered powerful insights into the selective pressures that operate on males; for many mammals, selection weighs heaviest on males in competition with other males for access to females. In addition, Bateman provided the framework that eventually led to an understanding of why males tend to compete for mates while females compete for resources. But the Bateman and the anisogamy paradigm also led us to overlook the full range of possible sources of variance in female reproductive success; not only variance arising from female–female competition over resources to translate into large gametes, but also variance arising from other factors as well. Not all females conceive. In some cases, such as marmosets, the presence of the dominant female suppresses ovulation in her subordinates. Some offspring, once conceived, are not carried to term. Among the factors leading to spontaneous abortion in baboons may be harassment by other females or the arrival of strange males (Mori & Dunbar, in press; Wasser & Barash, 1984). And of course, offspring once born need not survive. If born to a low-ranking toque macaque mother, a juvenile daughter may die of starvation, or if born to a mother chimp who for some reason is incapacitated, an offspring may be killed by a higher-ranking female. Having survived, a maturing female howler monkey may nevertheless find herself unable to join a breeding group and never have a chance to reproduce. A mother's condition, her competitive abilities, and her maternal skills are all very much at issue in the case of creatures such as primates. Yet, as amazing as it sounds, only relatively recently have primatologists begun to examine behaviors other than direct mother–infant interactions that affect the fates of infants (for elaboration see Hrdy, 1981; Small, 1984). Not the least among the variables affecting their survival is the role played by males, and the capacity of females to influence this male performance.

MALE INVOLVEMENT WITH INFANTS

Even for *Drosophila* it was a mistake to imagine that male investment never went further than chromosomes. Recent research makes it clear that, as in various butterflies and cockroaches, male fruitflies may sometimes transmit along with their sperm essential nutrients that otherwise would be in short supply (Markow & Ankney, 1984). When assumptions about minimal male involvement are extrapolated to species such as primates, however, far more than underestimation of male involvement is at stake. I would argue that it is not only ill-advised but impossible to understand primate breeding systems without taking into account the role of males in determining the survival or demise of infants.

There is probably no order of mammals in which male involvement with infants is more varied, more complex, or more crucial than among primates. About 10% of all mammalian genera exhibit some form of direct male care, that is the male carries the infant or provisions it. Among primates, however, the percentage of genera with direct, positive (if also sometimes infrequent) interactions between males and infants is roughly four times that, the highest figure reported for any order of mammals (Devra Kleiman & Malcolm, 1981; Vogt, 1984). Conversely, infanticide has been reported for over 15 different species of primates belonging to 8 genera and is probably widespread among apes and monkeys (Hausfater & Hrdy, 1984). Indeed, some male care is probably a direct outgrowth of the need by males to protect infants from other males (Busse & Hamilton, 1981). Yet, oddly, after two decades of intensive study of wild primates, we are only now beginning to scratch the surface of the rich interactions that exist between infants and adult males, which seem to have such critical repercussions for infant survival (see Hrdy, 1976; and especially, Taub, 1984a, 1984b). Effects of these relationships for infants after they grow up have rarely been investigated, although several researchers have recently suggested the possibility that fathers among gibbons and orangutans may play a role in helping their sons to set up or defend territories (MacKinnon, 1978; Tilson, 1981). These cases are of special importance because apart from intervention by brothers or by fathers in adopting an orphan (among gorillas and chimpanzees) direct, "maternal-like" care of infants by males is not typically seen among apes. But, the fact that parental investment by males does not take the same form as investment by females does not lessen its importance for offspring or its cost to the parent. My focus here is on primates, but I believe I could make many of the same points if I were a student of amphibians or fish in which male care is very common. One critical role of males is to protect immatures from distantly related conspecifics. It has long been assumed that one reason for male care among these species was the greater certainty of paternity permitted in species with external fertilization (i.e., the male can *know* which eggs he fertilized). But surely among these groups, as among primates, there has been selection on females to manipulate this situation.

The main exception to a general pattern of ignoring interactions between males and infants was of course the study of male care among monogamous primates. It has been known for over 200 years, ever since a zoologist-illustrator named George Edwards decided to watch the behavior of pet marmosets in a London garden, that among certain species of New World monkeys males contributed direct care for infants that equalled or exceeded that given by females (Edwards, 1758). Mothers among marmosets and tamarins typically give birth to twins, as often as twice a year, and to ease the female in her staggering reproductive burden the male carries the infant at all times except when the mother is actually suckling it. It was assumed that monogamy and male confidence of paternity was essential for the evolution of such care (Kleiman, 1977), and at the same time, it was assumed that monogamy among primates must be fairly rare (e.g., see Symons, 1979, or virtually any textbook on physical anthropology prior to 1981).

Recent findings, however, make it necessary to revise this picture. First of all, monogamy among primates turns out to be rather more frequent than previously believed (either obligate or facultative monogamy can be documented for some 17–20% of extant primates) and, second, male care turns out to be far more extensive than previously thought and not necessarily confined to monogamous species (Hrdy, 1981). Whereas, previously, it was assumed that monogamy and male certainty of paternity facilitated the evolution of male care, it now seems appropriate to consider the alternative possibility, whether the extraordinary capacity of male primates to look out for the fates of infants did not in some way pre-adapt members of this order for the sort of close, long-term relationships between males and females that, under some ecological circumstances, leads to monogamy! Either scenario could be true. The point is that on the basis of present knowledge there is no reason to view male care as a restricted or specialized phenomenon. In sum, though it remains true that mothers among virtually all primates devote more time and/or energy to rearing infants than do males, males nonetheless play a more varied and critical role in infant survival than is generally realized.

Male–infant interactions are weakly developed among prosimians, and in these primitive primates, male care more or less (but not completely) coincides with monogamy (Vogt, 1984). Direct male care occurs in 7 out of 17 genera, including one of the most primitive of all lemurs, the nest-building ruffed lemur (*Lemur variegatus*), where the male diligently tends the nest while the mother forages (personal communication from Patricia Wright). Among New World monkeys, 12 of 16 genera (Vogt, 1984) or, calculated differently, 50% of all species (Wright, 1984) exhibit direct male care, often with the male as the primary caretaker. That is, shortly after birth, an adult male—often with the help of various immatures in the group or other males—will take the infant, carry it (or them, in the frequent case of twins) on his back, share food with infants, either adult males or juveniles may catch beetles to feed them, or assist them by cracking the casing of tough fruit.

The role of males as primary caretakers for single (nontwin) infants is very richly developed among the night monkeys, *Aotus trivirgatus*. These small, monogamously mated, South American monkeys are the only nocturnal higher primate. Because of the difficulty in watching them, their behavior in the wild has gone virtually undocumented until detailed behavioral studies were undertaken by Patricia Wright using an image intensifier and other gear to allow her to work at night. Combining her observations of captive *Aotus* with field observations, a picture emerges in which the male is primary caretaker (in terms of carrying the infant) from the infant's first day of life, although the mother, of course, still is providing physiologically very costly milk. Based on captive observations, the mother carried the infant 33% of the time during the first week of life, the male 51%, and a juvenile group member 15%. In the wild, the infant was still being carried by the male at 4 months of age, although "weaning" tantrums were seen, as the male would try to push the infant off his back. By 5 months, the infant was relatively independent of either parent (Patricia Wright, 1984).

There is little question that there is an association between monogamy and extensive male care. Nevertheless, this does not mean that the evolution of male care is precluded by situations in which females mate with more than one male, as discussed for the case of savannah baboons.

Recent research on male–infant relations among baboons reveals that during their first week of life infant baboons at Amboseli spend about a third of their daylight hours within 5 feet of an adult male, often, but not always, a former sexual consort of the mother. This level of proximity was maintained throughout the first 7 weeks and then dropped sharply. At the same time, the amount of time infants spend in actual contact with an adult male, which is never much, is rising from 1% in the first week to 3% by the eighth week. During their first half-year of life, infants spend .5% of their time connected with an adult male, a low figure (Stein, 1984). Averaging together data from a number of different baboon field studies, David Taub calculates that a male–infant interaction takes place only about once every 19 hours (or, adjusting for the number of males in a multi-male troop, one interaction per male every 344 hours). However, Taub concurs with Busse and Hamilton (1981) and others, that the proximity of these males may be crucial for infant survival, particularly critical for discouraging attacks on the infant either by incoming males, unfamiliar with the infant's mother or, as suggested by Wasser (1983) for forestalling harassment by female troop members belonging to competing matrilines. That is, when the cost of care is fairly low (the male need only remain in the vicinity of the infant but can engage in other activities) and when it is rendered nonexclusively to several infants (e.g., to the offspring of each of the male's special female friends), male care certainly does occur in nonmonogamous systems. What is offered may not be "quantity" time, but it may well be "quality" time—"quality" in a very real sense: enhancing infant survival.

Yet, even these caveats can be dispensed with in the unusual case of the polyandrous tamarin species (*Saguinus fuscicollis*) studied by Goldizen. The female mates with several males and each of them subsequently helps rear the infant. Indeed, preliminary data from Goldizen's continuing research suggests that infants with several male caretakers are more likely to survive than infants born in small groups with only one adult male. Here, then, is both quality and quantity time, combined in a nonmonogamous breeding system, a system where males have a probability but no certainty of paternity. If we pause for a moment and consider the tamarin case from the male's point of view, the system Goldizen reports almost certainly derived initially from a monogamous one in which males were indeed caring for offspring likely to be their own. Only after such a system was established could a female have plausibly manipulated the situation to enlist the aid of two helpers.

Assuming that primate males do indeed remember the identity of past consorts and that they respond differentially to the offspring of familiar and unfamiliar females, females would derive obvious benefits from mating with more than one male. A researcher with this model in mind has quite different expectations about female behavior than one expecting females to save themselves in order to mate with the best available male. The resulting research questions will be very different.

THE ROLE OF WOMEN RESEARCHERS

When generalizations persist for decades after evidence invalidating them is also known, can there be much doubt that some bias was involved? We were predisposed to imagine males as ardent, females as coy; males as polygynists, females monandrous. How else could the *Drosophila* to primate extrapolation have entered modern evolutionary thinking unchallenged?

Assuming, then, this bias, a preconstituted reality in which males played central roles, what factors motivated researchers to revise invalid assumptions? What changes in the last decade brought about the new focus on female reproductive strategies and, with it, the recognition that certain assumptions and corollaries of the Bateman paradigm, and especially female monandry, were seriously limited and even, if applied universally, quite wrong.

The fact that there is relatively less intrasexual selection for mates among females does not mean reduced intrasexual competition or reduced selection among females in other spheres of activity. To understand male–male competition for mates is to understand only a small part of what leads to the evolution of particular primate breeding systems. We need also consider the many sources of variance in female reproductive success, including a whole range of female behaviors not directly related to "mothering" that may have repercussions on the fates of their infants.

Polyandrous mating with multiple males, mating with males when conception is not possible—what from the males' point of view might be termed "excessive" matings—can only be understood within this new framework, but it requires a whole new set of assumptions and research questions. As a result, sexual selection theory is currently in a state of flux; it is being rethought as actively as any area in evolutionary biology. What processes contributed to this destabilization of a long-held paradigm? And in particular, what led us to rethink the myth of the coy or monandrous female?

Improved methodologies and longer studies would not by themselves have led us to revise the myth of the coy female, simply because the relevant information about "female promiscuity" was already in hand long before researchers began to ask why females might be mating with more than one male. Indeed, at least one writer, working in a framework well outside of primatology and evolutionary biology, picked up on the reports of female promiscuity in baboons and chimpanzees at an early date (1966) and asked why it had evolved. This of course was the feminist psychiatrist Mary Jane Sherfey in her book, *The Evolution of Female Sexuality* (1973). Sherfey's vision of the "sexually insatiable" female primate was generally ignored by primatologists and biologists both because of her ideological perspective and because her standards of evidence were far from scientific. If her ideas were mentioned, it was typically with sarcasm and derision (Symons, 1979, pp. 76–77, 94, 262, 311). And, yet, it is important to note that however extreme her views (and scholarly balance was not Sherfey's strong point), they provided a valuable antidote to equally extreme ideas about universally coy females that were widely held by scientists within the academic mainstream of evolutionary biology. Elsewhere, I wrote about the various factors which caused us to recognize the importance of female dominance hierarchies in the lives of cercopithecine monkeys (Hrdy, 1984). Changes in methodology (e.g., focal animal sampling of all individuals in a group) and the emergence of long-term studies played critical roles in revising male-centered models of primate social organization. In that case as well, some of the relevant information was available long before we decided it was significant (e.g., the detailed Japanese studies indicating matrilineal inheritance of rank, Kawai, 1958; Kawamura, 1958). But, in the "coy female" case, I don't think that the duration of the studies or the field methods made as much difference as the particular research questions being asked. Ultimately, however, long-term studies are going to be very important for testing the various hypotheses to explain why females mate with multiple males.

New or better data alone did not change the framework in which we asked questions; rather, I believe, something motivational changed. Among the factors leading to a reevaluation of the myth of the coy female, the role of women researchers must be considered. That is, I seriously question whether it could have been just chance or just historical sequence that caused a small group of primatologists in the 1960s, who happened to be mostly male, to focus on male–male competition and on the number of matings males obtained, while a subsequent group of researchers, including many women (beginning in the 1970s), started to shift the focus to female behaviors having long-term consequences for the fates of infants (reviewed in Hrdy & Williams, 1983).

In this paper, I deliberately included first names whenever the work of a woman was cited. I did this to emphasize just how many women are currently working specifically in this area. Even a casual inspection reveals that women are disproportionately represented among primatologists compared to their representation in science generally. For example, in 1984, just over a third of the members (36%) of the American Society of Primatologists were women.* As we reconstruct the journey from Bateman (1948) to the recognition that the adjectives *coy* and *female* are something less than synonymous, it seems clear that the insights of women are implicated at every stage along the way and that their involvement exceeds their representation in the field. Having said this, I need to remind readers that as history my account here is biased by a conscious focus on contributions by women. A broader treatment would also have to describe the pioneering research on long-term male–female relations by T. M. Ransom and Robert Seyfarth and the extensive studies of male–infant relations by Mason, Mitchell, Redican, Stein, Taub,

*It should be noted however that membership in the ASP signals the *motivation* of women to join, since all one has to do is sign up and pay dues. Recognition and acceptance may be quite different. Contrast, for example, the position of women on editorial boards (4 of 40 on the *International Journal of Primatology* are women; 0 of 19 on the editorial board of the journal *Behavioral Ecology and Sociobiology*). When we examine the prestigious roster of *elected* fellows of the Animal Behavior Society for 1985, 1 of 62 is a woman. All 19 autobiographical chapters in *Leaders in the Study of Animal Behavior* are by men.

and others (see Taub, 1984a, 1984b, for reviews). I am acutely aware that my treatment here is biased both by my particular purpose (discussing the role of empathy by females for other females in causing us to revise old assumptions) and by my own involvement in the transition of primatology from the study of primate "behavior" to the study of primate "sociobiology." Hence, I leave to someone else the task of writing a balanced history of primatology in this period (e.g., see Alison Jolly, 1985).

The contributions of women researchers can be interpreted in several ways. Perhaps, women are simply better observers. As Louis Leakey used to say in an effort to justify his all-too-evident preference for women researchers, "You can send a man and a woman to church, but it is the woman who will be able to tell you what everyone had on" (personal communication, 1970). Or perhaps women are by temperament more pragmatic or more empirical, less open to theoretical bias. A difficulty with both ideas, of course, is that a few women were present in primatology in the 1960s, and both sexes participated in perpetuating myths about monkeys living in male-centered societies, where the primary activities of females had to do with mothering (e.g., Jane Goodall, 1971 or Phyllis Jay, 1963; but see Jane Lancaster, 1975; and Thelma Rowell, 1972, for exceptions). Women seemed just as vulnerable to bias as men.

If the presence of women was a constant but our ideas changed, perhaps, as Donna Haraway (1976) likes to remind us, the interpretations of primatologists simply mirror ideological phases in the history of the Western world. Indeed, it is disconcerting to note that primatologists are beginning to find politically motivated females and nurturing males at roughly the same time that a woman runs for vice president of the United States and Gary Trudeau starts to poke fun at "caring males" in his cartoons.

Or, perhaps, as Thelma Rowell (1984) suggested it was easier "for females to empathize with females, and . . . empathy is a covertly accepted aspect of primate studies" (p. 16). Perhaps, the insights were there all along but it took longer to challenge and correct male-centered paradigms because the perceptions of women fieldworkers lacked the authority of male theorists.

In *A Feeling for the Organism*, Evelyn Fox Keller (1983) hints at the possibility that women biologists may have some special sensibility concerning the creatures that they study, an ability to enter into the lives of their subjects—a suggestion that maize geneticist Barbara McClintock, the subject of her biography, would surely deny. Among other things, such a singular "gift" for women might be thought to confine women to particular areas of science or to diminish their accomplishments. That is, as primatologist Linda Fedigan wrote recently,

> I do admit to some misgivings about the wider implications of female empathy. Rowell may be correct about our sense of identification with other female primates, but I well remember my dismay when, having put many hours of effort into learning to identify the individual female monkeys of a large group,

> my ability was dismissed as being inherent in my sex by a respected and senior male colleague. (p. 308)

To put Fedigan's concern in perspective one needs to realize that in conversations with primatologists and, indeed, among ethologists generally, it is fairly commonplace to hear it said that women seem better able than men to learn to individually identify large numbers of animals. In a now legendary study, the seemingly incredible capacity of British ornithologist Dafila Scott to identify and remember hundreds of unmarked swans was tested by a male colleague. Indeed, it is occasionally suggested that the difficulty men have learning individuals is one reason why more men go into the ecological side of primatology.

Similarly, and I believe justifiably, women primatologists have worried about identifying too closely with the study of mothers and infants for fear that this area would become the "home economics" of primatology, a devalued women's domain within the discipline, or for fear that it would exacerbate the already common view that women study monkeys because it satisfies a deep-felt need to be around cuddly creatures.

Yet, suppose that there is some truth to the idea that women identified with same-sex subjects and allowed this identification to influence research focus? After all, isn't this what male primatologists, and many other ethologists as well, were doing throughout the 1960s and, occasionally, into the 1980s?

Even today, one can encounter lovely examples of what I call the *punch line phenomenon*, when a covert identification by researchers with same-sex individuals suddenly becomes overt in a last paragraph or emphatic comment. For example, in a seemingly impartial 1982 paper entitled, "Why Do Pied Flycatcher Females Mate with Already Mated Males?", the authors present data to show that females who mate with already mated males rear fewer offspring than female flycatchers who are the sole mates of males, regardless of the kind of territories they had to offer her. Surely, this modern, post-"coy female" paper, focused as it is upon the reproductive success of females, a paper essentially about female strategies, will not succumb to a male-centered perspective. Yet by the end of the paper, by some imperceptible process, the female has become object, the male protagonist: "Our conclusion is that polygamous pied flycatcher males deceive their secondary females" (p. 591) and the strategy works, according to the authors, because the females lack the time to check out whether the male already has a mate whose offspring he will invest in: "it pays for a pied flycatcher female to be fast rather than coy, and therefore *she* [italics mine] can be deceived. . . . "

My own work, before I began consciously to consider such matters, provides another example. The last line of *The Langurs of Abu: Female and Male Strategies of Reproduction* (1977), a book in which I scrupulously devoted equal space to both sexes, reads, "For generations, langur females have possessed the means to control their own destinies: caught in an evolutionary trap they have never been able to use them" (p. 309). I might as well have said *we*.

On a conversational level, few primatologists bother to deny this phenomenon. As a colleague remarked recently when the subject came up, "Of course I identify with them. I sometimes identify with female baboons more than I do with males of my own species." But why, we still need to ask, was the process of same-sex identification by women different in the 1970s and 1980s than in the early years of primatology?

I leave the general answers to such questions to social historians, who are more qualified than I to deal with them. At this point in the chapter, I abandon scholarship and attempt briefly to trace my own experiences as I remember them, particularly as they relate to the recognition of the active roles females were playing in the evolution of primate breeding systems.

REMINISCENCE

In 1970, as a first-year graduate student at Harvard, I began research on infanticidal behavior by males and ended, a decade later, almost entirely focused on the reproductive strategies of females. What processes were involved? Some months after starting my fieldwork in Rajasthan, India, I abandoned my original hypothesis (that infanticide was a response to crowding) and adopted an interpretation based on classical sexual selection theory: infanticide was an outcome of male–male competition for access to females. That is, males only killed infants when they (the males) invaded breeding units from outside; mothers whose infants were killed subsequently mated with the killer sooner than if the mothers had continued to lactate (Hrdy, 1974). By killing infants sired by other males, the usurpers increased their own opportunities to mate with fertile females.

The story was straightforward enough and in line with everything I had been taught at Harvard. But, there were loose ends, not the least of which was my growing emotional involvement with the plight of female langurs. Every 27 months, on average, some male was liable to show up and attempt to kill a female's infant, and increasingly, my identification was with the female victimized in this way, not with the male who, according to the sexual selection hypothesis, was thereby increasing his reproductive success. If infanticide really was an inherited male trait that could be elicited by particular conditions (as I believed was the case), why would females put up with this system? Why not refuse to breed with an infanticidal male and wait until a male without any genetic propensity for infanticide showed up? Consideration of this question led to many others related to the question of intrasexual competition among females generally (Hrdy, 1981).

First came an unconscious process of identification with the problems a female langur confronts followed by the formulation of conscious questions about how a female copes with them. This, in turn, led to the desire to collect data relevant to those questions. Once asked, the new questions and new observations forced reassessment of old assumptions and led to still more questions. Even events I had seen many times before (e.g., females leaving their troops to solicit extratroop males) raised questions as they never had before.

If it was really true that females did not benefit from additional matings, why were female langurs taking such risks to solicit males outside their troop? Why would already pregnant females solicit and mate with males? What influence might such behavior have for the eventual fate of the female's offspring? What were the main sources of variance in female reproductive success and what role did nonreproductive sexuality play in all this? Why is situation-dependent receptivity, as opposed to strictly defined cyclical receptivity or estrus, so richly developed in the order primates? Where did the idea of the coy female ever come from anyway? These are the questions that preoccupied me since 1977 and all of them grow out of an ability to imagine females as active strategists.

Yet, identification with same-sex individuals in another primate species may not be quite so simple as it sounds. This history of primatology suggests that the nature of this identification was changing over time as the self-image of women researchers also changed. In my own case, changes in the way I looked at female langurs were linked to a dawning awareness of male–female power relationships in my own life, though "dawning" perhaps overstates the case.

It would be difficult to explain to an audience of political activists how intelligent human beings could be as politically unaware as many field biologists and primatologists are. Almost by definition, we are people who lead isolated lives and, by and large, avoid joining groups or movements. In addition, I was the sole woman in my cohort, since I was the first woman graduate student my particular advisor had taken on and only toward the end of the 1970s did I begin to read anything by feminist scholars like Carolyn Heilbrun and Jean Baker Miller. Each step in understanding what, for example, might be meant by a term like *androcentric* was embarked upon very slowly and dimly, sometimes resentfully, as some savage on the fringe of civilization might awkwardly rediscover the wheel. When I did encounter feminist writings, I was often put off by the poor quality of the scholarship. Sherfey's book is a case in point: highly original insights were imbedded in what seemed to me a confused and often erroneous matrix. Nevertheless, the notion of "solidarity" with other women and, indeed, the possibility that female primates generally might confront shared problems was beginning to stir and to raise explicit questions about male–female relations in the animals I studied. That is, there were two (possibly more) interconnected processes: an identification with other females among monkeys taking place at roughly the same time as a change in my definition of women and my ability to identify and articulate the problems women confront.

Such an admission raises special problems for primatologists. My discipline has the choice of either dismissing me as a particularly subjective member of the tribe or else acknowledging that the tribe has some problems with objectivity. It is almost a cliche to mention now how male-biased the early animal behavior studies were (see Wasser, 1983). But, in the course of the last decade of revision, are we simply substituting a new set of biases for the old ones?

The feminist charge that most fields, including psychology, biology, and animal behavior, have been male-centered, is, I think, by now undeniable. Yet to me, the noteworthy and encouraging thing is how little resistance researchers in my own field have exhibited when biases are pointed out. Although I still sense in Britain a reluctance to admit that male bias was ever actually a problem, among primatologists in the United States it is now widely acknowledged, and this has to be a healthy sign. Indeed, in animal behavior and primatology, there has been something more like a small stampede by members of both sexes to study female reproductive strategies, as well as perhaps a rush to substitute a new set of biases for the old. (That is, among feminist scholars it is now permissible to say that males and females are different, provided one also stipulates that females are more cooperative, more nurturing, more supportive—not to mention equipped with unique moral sensibilities; among sociobiologists *kudos* accrue to the author of the most Machiavellian scenario conceivable.)

There are of course antidotes to the all-too-human element that plagues our efforts to study the natural world. Common sense in methodology is one. No one will ever again be permitted to make pronouncements about primate breeding systems after having studied only one sex or after watching only the conspicuous animals. A recognition of the sources of bias is another. If, for example, we suspect that identification with same-sex individuals goes on or that certain researchers identify with the dominant and others with the oppressed and so forth, we would do well to encourage multiple studies, restudies, and challenges to current theories by a broad array of observers. We would also do well to distinguish explicitly between what we know and what we know is only interpretation. But really (being generous) this is science as currently practiced: inefficient, biased, frustrating, replete with false starts and red herrings, but nevertheless responsive to criticism and self-correcting, and hence better than any of the other more unabashedly ideological programs currently being advocated.*

Acknowledgements—In the preface to her recent book *Mother Care: Other Care*, my colleague in behavioral biology Sandra Scarr (1984, p. xi) notes, "I wish I could thank all the wonderful graduate school professors who helped me to realize the joys of combining profession and motherhood; unfortunately there weren't any at Harvard in the early 1960s." A decade later, things

*Recent feminist programs advocating "conscious partiality" come to mind. If an unbiased knowledge is impossible, this argument runs, an explicitly biased, politically motivated approach is preferable to the illusion of impartial research.

at Harvard—at least in the biologically oriented part of Harvard that I encountered—had changed remarkably little. As I think back on those postgraduate years (my undergraduate experience at Harvard was a wonderful and very different story), I can not recall a single moment's fear of success, but what I do distinctly recall was the painful perception that there were professors and fellow students (no women in those years) who acted as if *they* feared that I might succeed. Intellectually, it was a tremendously exciting environment, filled with stimulating and occasionally inspirational teachers and coworkers. It was also an environment that was socially and psychologically hostile to the professional aspirations of women. But there were exceptions, exceptions made all the more significant because they were rare. In particular, I remain deeply grateful to Ed Wilson for encouragement and for unfailing behind-the-scenes support (probably the best kind) offered not only to me, but to other women ethologists both younger and older than myself.

Writing now from an ecological niche so benign as to cause me to wonder if perhaps my zeal as a feminist won't now lapse as a consequence, it is a pleasure to acknowledge discussions with Leo Berenstain, Daniel Hrdy, Jon Marks, David Olmsted, Peter Rodman, and Judy Stamps, who read and gave me detailed criticisms of this paper. I also thank Jeanne Altmann, Alison Jolly, Jane Lancaster, Linda Partridge, Joannie Silk, Meredith Small, Barbara Smuts, and the Pats, Whitten and Wright, for valuable discussion, and thank Ruth Bleier and Erik Eckholm for inciting me to think along the lines that I do at the end of this paper. Not least, I thank Nancy McLaughlin for her assistance in preparing the manuscript.

REFERENCES

Abernethy, V. (1978). Female hierarchy: An evolutionary perspective. In L. Tiger & H. Fowler (Eds.), *Female Hierarchies*. Chicago: Beresford Book Service.

Alatalo, R. V., Lundberg A., & Stahlbrandt, K. (1982). Why do pied flycatcher females mate with already mated males. *Animal Behaviour, 30*, 585–593.

Altmann, J. (1980). *Baboon mothers and infants*. Cambridge: Harvard University Press.

Altmann, S. (Ed.). (1965). *Japanese monkeys: A collection of translations*. Edmonton, Canada: The editor.

Andelman, S. (forthcoming). Concealed ovulation and prolonged receptivity in vervet monkeys (*Cercopithecus aethiops*).

Bateman, A. J. (1948). Intra-sexual selection in drosophila. *Heredity, 2*, 349–368.

Benshoof, L., & Thornhill, R. (1979). The evolution of monogamy and concealed ovulation in humans. *Journal of Biological Structures, 2*, 95–106.

Bleier, R. (1984). *Science and gender*. Elmsford, NY: Pergamon.

Bray, O. E., Kennelly, J. J., & Guarino, J. L. (1975). Fertility of eggs produced on territories of vasectomized red-winged blackbirds. *Wilson Bulletin, 87*, no. 2, 187–195.

Burley, N., & Symanski, R. (1981). Women without: An evolutionary perspective on prostitution. In *The immoral landscape: Female prostitution in Western societies*. Toronto: Butterworth.

Busse, C., & Hamilton, W. J., III. (1981). Infant carrying by male chacma baboons. *Science, 212*, 1281–1283.

Clutton-Brock, T. H., & Harvey, P. (1976). Evolutionary rules and primate societies. In P. P. G. Bateson & R. A. Hinde (Eds.), *Growing points in ethology*. Cambridge: Cambridge University Press.

Cords, M. (1984). Mating patterns and social structure in redtail monkeys (*Cercopithecus ascanius*). *Zeitschrift für Tierpsychologie, 64*, 313–329.

Cronin, C. (1980). Dominance relations and females. In D. R. Omark, F. F. Strayer, and D. G. Freeman (Eds.), *Dominance relations*. New York: Garland Press.

Daly, M., & Wilson, M. (1983). *Sex, evolution and behavior*. Boston: Willard Grant Press.

Darwin, C. (1871). *The descent of man and selection in relation to sex* (1887 edition). New York: D. Appleton and Co.
Davies, E. M., & Boersma, P. D. (1984). Why lionesses copulate with more than one male. *The American Naturalist, 123*, no. 5, 594–611.
Davies, N. B., & Lundberg, A. (1984). Food distribution and a variable mating system in the dunnock, *Prunella modularis. Journal of Animal Ecology, 53*, 895–912.
Dawkins, R. (1976). *The selfish gene*. Oxford: Oxford University Press.
DeVore, I. (Ed.). (1965). *Primate behavior*. New York: Holt, Rinehart and Winston.
Diamond, J. (1984). Theory and practice of extramarital sex. *Nature, 312*, 196.
Dunbar, R. (1984). *Reproductive decisions: An economic analysis of gelada baboon social strategies*. Princeton, NJ: Princeton University Press.
Eaton, R. (Ed.). (1976). *The world's cats II*. Seattle, WA: Feline Research Group, Woodland Park Zoo.
Edwards, G. (1758). *Gleanings of Natural History* (Vol. 5). London: College of Physicians.
Fedigan, L. (1984). Sex ratios and sex differences in primatology (book review of *Female primates*). *American Journal of Primatology, 7*, 305–308.
Freedman, D. (1979). *Human sociobiology: A holistic approach*. New York: The Free Press.
Fujioka, M., & Tamagishi, S. (1981). Extramarital and pair copulations in the cattle egret. *Auk, 98*, 134–144.
Gladstone, D. (1979). Promiscuity in monogamous colonial birds. *The American Naturalist, 114*, no. 4, 545–557.
Goldizen, A. W., & Terborgh, J. (in press). Cooperative polyandry and helping behavior in saddle-backed tamarins (*Saguinus fuscicollis*). Proceedings of the IXth Congress of the International Primatological Society. Cambridge: Cambridge University Press.
Goodall, J. (1971). *In the shadow of man*. Boston: Houghton Mifflin.
Haraway, D. (1976). The contest for primate nature: Daughters of man-the-hunter in the field. In M. Kann (Ed.), *The future of American democracy: Views from the left*. Philadelphia, PA: Temple University Press.
Hartung, J. (in press). Matrilineal inheritance: New theory and analysis. *The Behavioral and Brain Sciences*.
Hausfater, G. (1975). Dominance and reproduction in baboons (*Papio cynocephalus*). *Contributions to Primatology* (Vol. 7). Basel, Switzerland: S. Karger.
Hausfater, G., & Hrdy, S. B. (Eds.). (1984). *Infanticide: Comparative and evolutionary perspectives*. New York: Aldine.
Hrdy, D. B. (1979). Integrated field study of the behavior, genetics and diseases of the Hanuman langur in Rajasthan, India. Proposal submitted to the National Science Foundation.
Hrdy, S. B. (1974). Male-male competition and infanticide among the langurs (*Presbytis entellus*) of Abu, Rajasthan. *Folia Primatologica, 22*, 19–58.
Hrdy, S. B. (1976). The care and exploitation of nonhuman primates by conspecifics other than the mother. *Advances in the Study of Behavior, VI*, 101–158.
Hrdy, S. B. (1977). *The langurs of Abu: Female and male strategies of reproduction*. Cambridge: Harvard University Press.
Hrdy, S. B. (1979). Infanticide among animals: A review, classification and examination of the implications for the reproductive strategies of females. *Ethology and Sociobiology, 1*, 3–40.
Hrdy, S. B. (1981). *The woman that never evolved*. Cambridge: Harvard University Press.

Hrdy, S. B. (1984). Introduction: Female reproductive strategies. In M. Small, (Ed.), *Female primates: Studies by women primatologists*. New York: Alan Liss.
Hrdy, S. B., Hrdy, D. B., & Bishop, J. (1977). *Stolen copulations*. 16 mm color film. Peabody Museum.
Hrdy, S. B., & Whitten, P. (1986). The patterning of sexual activity. In D. Cheney, R. Seyfarth, B. Smuts, R. Wrangham, & T. Struhsaker (Eds.), *Primate societies*. Chicago: University of Chicago Press.
Hrdy, S. B., & Williams, G. C. (1983). Behavioral biology and the double standard. In S. K. Wasser (Ed.), *Social behavior of female vertebrates*. New York: Academic Press.
Jay, P. (1963). The female primate. In S. Farber & R. Wilson (Eds.), *The potential of woman*. New York: McGraw-Hill.
Jolly, A. (1985). *The evolution of primate behavior*. New York: Macmillan.
Kawai, M. (1958). On the system of social ranks in a natural troop of Japanese monkeys: I. Basic rank and dependent rank. *Primates, 1-2*, 111-130.
Kawamura, S. (1958). Matriarchal social ranks in the Minoo-B troop: A study of the rank system of Japanese monkeys. *Primates, 1-2*, 149-156.
Keller, E. F. (1983). *A feeling for the organism: The life and work of Barbara McClintock*. New York: W. H. Freeman.
Kleiman, D. (1977). Monogamy in mammals. *Quarterly Review of Biology, 52*, 39-69.
Kleiman, D., & Malcolm, J. (1981). The evolution of male parental investment in mammals. In D. J. Gubernick & P. H. Klopfer (Eds.), *Parental care in mammals*. New York: Plenum Press.
Koyama, N. (1967). On dominance rank and kinship of a wild Japanese monkey in Arashiyama. *Primates, 8*, 189-216.
Lamb, M. (1984). Observational studies of father-child relationships in humans. In D. Taub (Ed.), *Primate paternalism*. New York: Van Nostrand Reinhold.
Lancaster, J. (1975). *Primate behavior and the emergence of human culture*. New York: Holt, Rinehart and Winston.
Lott, D. (1981). Sexual behavior and intersexual strategies in American Bison. *Zeitschrift für Tierpsychologie, 56*, 97-114.
Lumpkin, S. (1983). Female manipulation of male avoidance of cuckoldry behavior in the ring dove. In S. C. Wasser (Ed.), *The social behavior of female vertebrates*. New York: Academic Press.
MacKinnon, J. (1978). *The ape within us*. New York: Holt, Rinehart and Winston.
Markow, T. A., & Ankney, P. F. (1984). *Drosophila* males contribute to oogenesis in a multiple mating species. *Nature, 224*, 302-303.
Moore, J. (1985). Demography and sociality in primates. Doctoral dissertation, Harvard University. Cambridge.
Mori, U., & Dunbar, R.I.M. (in press). Changes in the reproductive condition of female gelada baboons following the takeover of one-male units. *Zeitschrift für Tierpsychologie*.
Morris, S. (1979, August). Darwin and the double standard. *Playboy Magazine*.
Parker, G. A. (1970). Sperm competition and its evolutionary consequences in the insects. *Biological Review, 45*, 525-567.
Ransom, T., & Ransom, B. (1971). Adult-male-infant interactions among baboons (*Papio anubis*). *Folia Primatologica, 16*, 179-195.
Rowell, T. (1972). *Social behaviour of monkeys*. Baltimore, MD: Penguin Books.
Rowell, T. (1984). Introduction: Mothers, infants and adolescents. In M. Small (Ed.), *Female primates*. New York: Alan Liss.

Sayers, J. (1982). *Biological politics*. London: Tavistock.
Scarr, S. (1984). *Mother care: Other care*. New York: Basic Books.
Seyfarth, R. (1978). Social relationships between adult male and female baboons, part 2: Behavior throughout the female reproductive cycle. *Behaviour, 64*, nos. 3-4, 227-247.
Shaw, E., & Darling, J. (1985). *Female strategies*. New York: Walker.
Sherfey, M. J. (1973). *The evolution of female sexuality* (first published 1966). New York: Vintage Books.
Small, M. (Ed.). (1984). *Female primates*. New York: Alan Liss.
Small, M. (Forthcoming). Primate female sexual behavior and conception: Is there really sperm to spare?
Smith, J. M. (1984). Optimization theory in evolution. In E. Sober (Ed.), *Conceptual issues in evolutionary biology*. Cambridge, MA: The M.I.T. Press.
Smith, R. (1984). Sperm competition. In *Sperm competition and the evolution of animal mating systems*. New York: Academic Press.
Smuts, B. B. (1985). *Sex and friendship in baboons*. New York: Aldine Publishing Co.
Stacey, P. B. (1982). Female promiscuity and male reproductive success in social birds and mammals. *The American Naturalist, 120*, no. 1, 51-64.
Stein, D. (1981). The nature and function of social interactions between infant and adult male yellow baboons (*Papio cynocephalus*). Doctoral dissertation, University of Chicago.
Stein, D. (1984). Ontogeny of infant-adult male relationships during the first year of life for yellow baboons (*Papio cynocephalus*). In D. Taub (Ed.), *Primate paternalism*. New York: Van Nostrand Reinhold.
Symons, D. (1979). *The evolution of human sexuality*. Oxford: Oxford University Press.
Taub, D. (1980). Female choice and mating strategies among wild Barbary macaques (*Macaca sylvana*). In D. Lindburg (Ed.), *The macaques*. New York: Van Nostrand Reinhold.
Taub, D. (1984a). Male-infant interactions in baboons and macaques: A critique and reevaluation. Paper presented at the American Zoological Society Meetings, Philadelphia, PA.
Taub, D. (1984b). *Primate paternalism*. New York: Van Nostrand Reinhold.
Tiger, L. (1977). The possible biological origins of sexual discrimination. In D. W. Brothwell (Ed.), *Biosocial man*. London: The Eugenics Society.
Tilson, R. (1981). Family formation strategies of Kloss' gibbons. *Folia Primatologica, 35*, 259-287.
Trivers, R. L. (1972). Parental investment and sexual selection. In B. Campbell (Ed.), *Sexual selection and the descent of man*. Chicago: Aldine.
Trivers, R. L. (1985). *Social evolution*. Menlo Park, CA: Benjamin/Cummings.
Tsingalia, H. M., & Rowell, T. E. (1984). The behaviour of adult male blue monkeys. *Zeitschrift für Tierpsychologie, 64*, 253-268.
Turner, M. E., & Anderson, W. W. (1983). Multiple mating and female fitness in *Drosophilia pseudoobscura. Evolution, 37*, no. 4, 714-723.
Tutin, C. (1975). Sexual behaviour and mating patterns in a community of wild chimpanzees (*Pan troglodytes schweinfurthii*). Doctoral dissertation submitted to the University of Edinburgh, Edinburgh.
Van den Berghe, E. (1984). Female competition, parental care, and reproductive success in salmon. Paper presented at Animal Behavior Society Meetings, Cheney, Washington, August 13-17.

Vogt, J. (1984). Interactions between adult males and infants in prosimians and New World monkeys. In D. Taub (Ed.), *Primate paternalism*. New York: Van Nostrand Reinhold.

Wasser, S. C. (Ed.). (1983). *The social behavior of female vertebrates*. New York: Academic Press.

Wasser, S. C., & Barash, D. (1984). Reproductive suppression among female mammals. *Quarterly Review of Biology*, 513–538.

Williams, G. C. (1966). *Adaptation and natural selection*. Princeton, NJ: Princeton University Press.

Williams, G. C. (1975). *Sex and evolution*. Princeton, NJ: Princeton University Press.

Wilson, E. O. (1978). *On human nature*. Cambridge: Harvard University Press.

Wirtz, P. (1983). Multiple copulations in the Waterbuck. *Zeitschrift für Tierpsychologie, 61*, 78–82.

Wright, P. (1984). Biparental care in *Aotus trivirgatus* and *Callicebus molloch*. In M. Small (Ed.), *Female primates*. New York: Alan Liss.

Zuckerman, Sir S. (1932). *The social life of monkeys and apes*. London: Butler and Turner, Ltd.

Chapter 7

Sex Differences Research: Science or Belief?

Ruth Bleier

Research on presumed sex differences in cognitive abilities is an area of the natural sciences in need of the drastic revision that feminists effected in the field of primatology. This field is fraught with unexamined or untested assumptions, with inconclusive and contradictory findings and misleading interpretations that become incorporated into belief systems called *theories*, and with the reckless use of language designed to appeal to the news media and a reading public highly susceptible to scientific pronouncements, especially those that confirm common beliefs.

Over the past two decades, an enormous scientific literature on sex differences* has accumulated in a variety of biological and social scientific disciplines: behavioral endocrinology, neuroanatomy, psychology, primatology, political science, and others. Much of the literature attempts to explain observed differences in social roles or socially recognized achievements between women and men in our society. Neuroscientists have looked for the biological bases for these differences in prenatal hormonal effects on the developing brain and in differences in brain hemispheric lateralization.

I believe that the heightened interest in biological sex differences either to explain or justify the myriad forms of gender asymmetries is not unrelated to the social–political context of the 1970s and 1980s, when the women's movement has forced into public scrutiny and policy questions of inequalities in employment and education and in legal and social status. This sensitivity of science to social events and values is not new. One of the most compelling

*I usually use the term *sex differences* in this paper, since that is the generally accepted name given to this field of research in the various disciplines. The phenomena being studied, however, are *gender* differences, if we accept that sex refers to categories defined biologically by genitalia and reproductive organs. See Haraway (chapter 5), however, who warns that it is important not to make the mistake of thinking that sex is given, natural, and biological and that only gender is constructed and social.

studies of social influences on scientists and their beliefs in the form of scientific pronouncements on race and intelligence is that of William Provine (1973). As he wrote, in the wake of the Civil War, the freeing of slaves in the United States, and the colonization of European nations of Africa and other parts of the world, race-related social problems were prominent. Biologists concurred with the dominant view of North American and European whites that the problems resulted from the mental inferiority of nonwhite races, and they were strong advocates of eugenics through the latter part of the 19th and early 20th century. At the turn of the century, geneticists were able to use Mendelian genetics and results from plant crossing to support their beliefs about the deleterious effects of race crossing. But then, in the course of two critical decades between 1930 and 1950, geneticists in England and the United States reversed in two steps their published beliefs on the effects of race crossing. First, they moved from condemnation of race crossing to an agnostic neutrality on the subject. Then, following the second World War with its Nazi ideology and projects for racial purity, they moved further from agnostic neutrality to the belief that race crossings were at worst biologically harmless. This reversal occurred with little new data from studies of actual human race crossing.

Provine gave a warning that is very appropriate today.

> I am not condemning geneticists because social and political factors have influenced their scientific conclusions about race crossing and race differences. It is necessary and natural that changing social attitudes will influence areas of biology where little is known and the conclusions are possibly socially explosive. The real danger is not that biology changes with society, but that the public expects biology to provide the objective truth apart from social influences. (Provine, 1973, p. 796)

That social values and beliefs infiltrate the work of scientists is hardly surprising. In fact, it would be naive to believe that scientists, unlike other human beings, are unaffected in their work by their life history of experiences and perceptions and by the values and realities of their culture. This is a subject that has been explored extensively by philosophers and historians of science, sociologists of knowledge, and scientists themselves (Lewontin, Rose, & Kamin, 1984; Mendelsohn, Weingart, & Whitley, 1977). The brain has frequently been the battle site in controversies over sex or race differences.

Today, we can see clearly the biases of some of the most reputable scientists working in the middle and late 19th century, a period of turmoil in regard to slavery and women's rights, who found female and "Negro" brains to be inferior and underdeveloped (Fee, 1979). Stephen Jay Gould has shown how the most prominent brain scientists of the period, obsessed with numbers as indicators of scientific rigor, used craniometry to "confirm all the common prejudices of comfortable white males—that blacks, women, and poor people occupy their subordinate roles by the harsh dictates of nature" (1981, p. 74).

G. Le Bon, whom Gould calls the chief misogynist of Broca's school, wrote in 1879:

> In the most intelligent races, as among the Parisians, there are a large number of women whose brains are closer in size to those of gorillas than to the most developed male brains. This inferiority is so obvious that no one can contest it for a moment; only its degree is worth discussion. (Gould, 1981, pp. 104–105)

It is, however, difficult to see clearly something that is happening in science *today* and to believe that *today's* social values and problems can affect the questions that scientists find interesting to ask, the methods used, the interpretations made of data, and the alternative interpretations not considered.

Scientists who have made important contributions to their fields have shown serious suspensions of critical judgment in interpretations of their own and others' data to make them fit what has become a ruling paradigm of the 1970s and 1980s. That paradigm is the assumption that significant cognitive sex differences exist and that these differences may be explained by looking for biological sex differences in the development, structure, and functioning of the brain. Support of the paradigm involves ignoring the obvious and the known (the complexity and malleability of human development and ability) in favor of a constructed reality (of biologically based gender differences), legitimized by an elaborate network of interdependent hypotheses, as I shall demonstrate. Few of the hypotheses and assumptions have independent scientific support but together, supported by each other, they create the illusion of a structure with weight, consistency, conviction, and reason. Using this paradigm, scientists have been making increasing numbers of unsubstantiated conjectures that are then taken up by other scientists as confirming evidence for their own conjectures, a process that repeats itself like an infinite series of images produced by two facing mirrors.

The troublesome truth is that no single part of the paradigm is known to be descriptive of scientifically verifiable reality as we know it today. It is the goal of this paper to illustrate this by analyzing several well-known studies in three research areas prominently featured in the sex-differences literature: aggressivity, brain hemispheric asymmetries, and mathematical/visuospatial ability.*

*Space limitations preclude a more complete analysis. See R. Bleier, *Science and Gender: A Critique of Biology and Its Theories on Women* (New York: Pergamon, 1984) for a lengthier discussion of some issues raised in this paper.

AGGRESSIVITY

Following experimental work in rodents demonstrating a relationship between pre- and perinatal hormonal levels and the incidence of fighting behaviors in the adult (Conner & Levine, 1969; Edwards, 1969), there has been interest in the possible relationship between androgen levels and "aggressivity" in humans. Possibly, the most widely cited studies in the literature supporting sex differences in human aggressivity are those on "tomboyism" in girls exposed as fetuses to high levels of androgens (Ehrhardt, Epstein, & Money, 1968; Ehrhardt & Money, 1967). The 25 fetally androgenized teenagers that constituted the subject population for these two studies were born with masculinized genitalia, usually resembling a penis and scrotum closely enough that some were considered to be boys for a while and, in all but one case, requiring reconstruction of the clitoris, labia, and vagina. The authors found a higher level of "tomboyism" in the subjects than in controls. By tomboyism, the authors meant a preference for outdoor play and athletics, for boys as playmates, and for boys' toys; little interest in dolls or in infant care; preference for functional clothes (pants rather than dresses); an interest in career equal to or greater than that in marriage. The tomboyism reported in these studies has been widely and uncritically interpreted as reflecting "masculinization" of the developing brain and as being a measure of aggressivity, induced by androgens.

It is a measure of the degree to which "membership in a culture blinds us to the constructed nature of that culture's reality" (Kessler & McKenna, 1978) that the authors and subsequent scientists accept at face value the idea of tomboyism as an index of a characteristic called *masculinity*, presumed to be as objective and innate a human feature as height and eye color. Yet "masculinity" is a gender characteristic and, as such, culturally, not biologically, constructed (though, indeed, the construction proceeds from a gender attribution established at birth on the basis of biological features: the presence of a penis and scrotum). The attribution of masculinity or maleness to physical activity, athleticism, and freedom of body movement is an anachronistic 19th century, white EuroAmerican value judgment, not a scientific truth. Its necessary complement is the 19th century prescription of a presumably innate characteristic called *femininity*: restriction of body movements, athletic incompetence, passivity, and preoccupation with maternalism and nurturant activities. This is an example of psychology's turning "the nineteenth-century nurturing imperative into a twentieth-century factual, female attribute" (Westkott, 1984).

This concept of tomboyism further implies that if little girls do not conform to the cultural stereotype, something must be wrong with them: disordered hormones or chromosomes or gender identity. Such value judgments form the unexpressed assumptions for the research, and they affect what and how data are collected and what interpretations are made of the data. The finished work then provides a body of "knowledge" that places the stamp of *science* on a set of unexamined social values and judgments concerning gender. Thus, science and culture together structure gender and gender characteristics in and by the very process of their explorations of presumed biological origins of socially constructed gender differences.

Furthermore, the social science and biological literature on sex differences in aggressivity has uncritically supported the belief that Ehrhardt and Money demonstrated a cause and effect relationship between *in utero* effects of androgens on the brain and subsequent behaviors. One need only read the original reports, described below, to see that the authors did not demonstrate an effect of androgens on the developing brain in the production of tomboyism, and that they themselves concluded in their initial reports that this was only one of several possible interpretations of their findings. That this tentative hypothesis was rapidly incorporated into the literature as a demonstrated fact had more to do with the social and political climate of the 1970s and 1980s than with rigorous scientific inquiry.

The first study was of 10 girls with progestin-induced fetal androgenization, the second was of 15 girls with the adrenogenital syndrome (AGS). In their first paper, Ehrhardt and Money raised the "question of whether tomboyishness may not be a frequent characteristic in the development of middle-class suburban and rural girls who have both the space and tradition of the outdoor life" (p. 96). They concluded that "It will require more than ten cases and better control of at least the socioeconomic variable before one can answer with confidence the question of the extent to which prenatal hormones can affect subsequent behavior" (p. 98). In their second paper, they wrote

> It is not possible to estimate on the basis of present data, whether individual differences in degrees of tomboyism may have reflected differences in parental attitude. Each parent knew of the child's genital masculinization at birth. This knowledge may have insidiously influenced their expectancies and reactions regarding the child's behavioral development and interests . . . (Ehrhardt, Epstein, & Money, 1968, p. 166)

Yet, on the basis of the same data on the same 25 patients, and ignoring their cautious reservations and qualifications at the time of original publication, Ehrhardt and Money wrote a number of subsequent articles, chapters, and a book in which they became ever more assertive that the results demonstrated the behavioral effects of androgenization of the fetal female brain. In their book, written 4 years later, they wrote, "The most likely hypothesis to explain the various features of tomboyism in fetal masculinized genetic females is that their tomboyism is a sequel to a masculinizing effect on the fetal brain" (Money & Ehrhardt, 1972, p. 103).

Following a subsequent similar study of 17 female AGS patients in 1974, Ehrhardt and another colleague concluded that their findings "suggest strongly that it is the fetal exposure to androgens that contribute to the typical profile of behavior exhibited by AGS females" (Ehrhardt and Baker, 1974, p. 48). On the basis of the same body of work, but 5 years later, Ehrhardt and another colleague concluded that the "effects of prenatal androgens *have been established* [emphasis added] for the sex-dimorphic behavior clusters" (Ehrhardt and Meyer-Bahlburg, 1979, p. 41).

Aside from the reservations originally expressed by the authors themselves, there are a number of other reasons for questioning any claim that tomboy behaviors (i.e., aggressivity) were demonstrated to be a product of androgen effects on the fetal brain. I shall mention only a few here. The first is that the authors did not consider the effects on a child's choices and identifications of knowing that she has or had a penis and scrotum, the one accepted sign of being a boy. The first plastic surgery was not done in some cases until 3½- or even 7½-years of age, and usually further vaginal surgery was necessary at the time of adolescence. For both themselves and their parents, surely the ambiguity of their genitalia and their situation in this extremely gender-polarized society would have some effect on their preferences and choices in play, playmates, clothes, and attitudes toward motherhood and career. When gender becomes so fragile and arbitrary a concept that it can be changed by plastic surgery, perhaps parents and children alike recognize the arbitrariness and limitations of gender-dictated choices about behaviors, dress, play, and life's work. That is, it would seem arbitrary indeed, under the circumstances, not to permit or encourage boy-characteristic activities, along with "male" as well as "female" life goals, in a girl born with a boy's genitalia, the one usual absolute indicator for gender assignment at birth. It also seems relevant that, in the first study of 10 girls, the only girl who preferred dolls, showed no preference for outdoor play or boys' toys or clothes, and did not consider herself a tomboy was the *only* one without masculinized genitalia. While Ehrhardt and Money noted the possible importance of parental attitudes and family traditions, they did not formally explore the attitudes or study the siblings of the patients. Yet their anecdotal comments are revealing. They recorded, in their first study of 10 patients, that one mother considered herself a tomboy as a child and reported that *both* her daughters, the patient and her unaffected sister, were tomboys. They added that it "was anecdotally evident . . . that some of the sisters of index cases were tomboyish . . . " (Ehrhardt & Money, 1967, p. 97). Since some (number unspecified) of the 10 patients had *no* sisters, and we know that some (number unspecified) of the patients had sisters who were tomboys, it seems rather clear that tomboyish behaviors may have been characteristic rather than exceptional in this small sample of families. The data, as they are reported or implied, do not easily invite the elaborate, and scientifically dubious, explanation of an androgen effect on the developing fetal brain.

HEMISPHERIC LATERALIZATION AND COGNITIVE FUNCTIONING

Recent years have seen a heightened interest in finding sex differences in brain structure and function to explain presumed sex differences in cognitive ability. The focus has been mainly on the question of hemispheric lateralization in cognitive processing. The predominant theory is that women tend to use the left hemisphere in addition to the right hemisphere in the processing of visuospatial information, whereas men rely more exclusively on the right hemisphere in visuospatial processing. Thus, women are said to be less lateralized or less specialized (with regard to the hemispheres) than men in processing visuospatial information, though more lateralized to the left hemisphere for processing verbal information.

It is, first of all, noteworthy that the majority of studies in this area are flawed for one reason or another and that there is no agreement among them on the question of sex differences in lateralization of visuospatial processing, as five exhaustive reviews of the literature have documented (Alper, 1985; Caplan, MacPherson, & Tobin, 1985; Fairweather, 1976; Kimball, 1981; McGlone, 1980). Just as many published studies find no sex differences in lateralization of cognitive processing as do find them, and probably many that find no sex differences are not published because they are of no interest. Variability within each sex is greater than variability between them. When sex differences are found, they often have a very weak if any statistical significance, and the statistical differences that exist are between *mean* scores of the two groups tested whose scores mainly overlap. Furthermore, results obtained in tests of lateralization of cognitive functioning (and therefore any sex differences that are found) are strongly influenced (i.e., enhanced, negated, or reversed) by such uncontrolled or uncontrollable factors as age, test procedures, task difficulty, information-processing strategies by the subject, practice doing the task, attention, motivation, memory duration, and general aptitude. The fragile nature of the findings clearly do not support any theory about cognitive sex differences, as three of the review articles noted above emphasize.

In fact, even the review paper most widely quoted as supporting sex differences in hemispheric lateralization concludes, "Thus, one must not overlook perhaps the most obvious conclusion, which is that basic patterns of male and female brain asymmetry seem to be more similar than they are different" (McGlone, 1980, p. 226). Yet, while admitting that no conclusions can be drawn from the inadequate database in the literature, McGlone clearly supports the notion of sex differences in lateralization. One may well wonder why. In his commentary on McGlone's paper, Marcel Kinsbourne, one of the leaders in this field, wrote

> We have seen that the evidence for sex-differential lateralizaton fails to convince on logical, methodological, and empirical grounds. Is that surprising? Not

all the points made in this critique are subtle, and some at least must be obvious to anyone in the field. Why then do reputable investigators persist in ignoring them? Because the study of sex differences is not like the rest of psychology. Under pressure from the gathering momentum of feminism, and perhaps in backlash to it, many investigators seem determined to discover that men and women "really" are different. It seems that if sex differences (e.g., in lateralization) do not exist, then they have to be invented. (Kinsbourne, 1980, p. 242)

Reviewing the history of research in sex differences in cognition into the mid-1970s, Hugh Fairweather (1976) wrote, "What had before been a possibility at best slenderly evidenced, was widely taken for fact; and 'fact' hardened into a 'biological' dogma" (p. 233). He concluded his review by saying

It must be stressed, finally, that the majority of studies reviewed here and elsewhere are both ill-thought and ill-performed. Whilst in other circumstances this may be regarded as the occupational hazard of the scientific enterprise, here such complacency is compounded by the social loadings placed upon these kinds of results. . . . We cannot pretend that we are testing a theory of sex differences, since at present none can exist. (p. 267)

Yet, so far as I can tell, these devastating criticisms by two leaders in the field of cognitive sex differences and lateralization have done nothing to stem the flood of research on cognitive sex differences.

But another serious interpretive problem is that, even if sex differences in hemispheric lateralization of visuospatial function *were* clearly demonstrated, there is no evidence of any correlation between hemispheric lateralization and visuospatial *ability*. One would never know this from the literature, however. One reason is that language is used carelessly and conceals this lack of knowledge. Instead of the term hemispheric *lateralization*, often the term hemispheric *specialization* is used. Men are specialized to the right hemisphere, and we know that to be specialized is to be better.

There are studies that find sex differences in lateralization but no sex differences in spatial abilities in the same test groups, studies that find a sex difference in spatial abilities but no sex difference in lateralization, and studies that find no sex differences in either lateralization or ability (Kimball, 1981). The assumption is that the (questionable) demonstration of right hemispheric lateralization of visuospatial processing in males accounts for their presumed superiority in visuospatial skills. But no independent evidence supports this assumption. It is instead a product of circular reasoning: men are superior in visuospatial skills because their right hemispheres are specialized for visuospatial cognitive processing; we know that right hemispheric specialization provides superior visuospatial skills because men have better visuospatial skills than women, who use both hemispheres for visuospatial processing. To put this another way, if it is true that women use both hemispheres for processing visuospatial information, there is no intrinsic reason to believe that that situation makes for inferior rather than superior visuospatial processing.

The implications of this area of research are even more problematic when visuospatial processing and abilities are assumed to be synonymous with mathematical ability. While it is reasonable to assume an association between visuospatial processing and *geometrical* problem-solving, much of mathematics involves logic, analysis, and reasoning, considered to be predominantly left hemispheric functions. This introduces yet another problem for the dominant theory, which claims women to be more left hemispherically specialized than men.

Some recent studies purport to add evidence for this dominant belief in sex differences in lateralization and cognitive processing.

INTERHEMISPHERIC TRANSFER OF INFORMATION

A recent study (De Lacoste-Utamsing & Holloway, 1982) reports that the splenium (the caudal part of the corpus callosum, which carries nerve fibers connecting the two hemispheres) was larger and more bulbous in five female brains than in nine male brains obtained at autopsy. No mention is made of how the particular 14 brains were selected for measurement nor of the ages or circumstances of death in these subjects, factors of possible relevance to any measurements of the brain. The authors state that a large sample size is needed before further interpretation of these results can be made, but they nevertheless offer their interpretation of the significance of the findings and, furthermore, claim that their finding "has widespread implications for students of human evolution and comparative neuroanatomy, as well as for neuropsychologists in search of an anatomical basis for possible gender differences in the degree of cerebral lateralization" (p. 1431). A modest claim for a study based on 14 brains and reporting a finding with no statistical significance ($p = 0.08$). After citing some evidence that the splenium is the site of the interhemispheric transfer of visual information, the authors write

> If we are to believe that a larger splenium implies a larger number of fibers interconnecting cortical areas and that the number of interhemispheric fibers correlates inversely with lateralization of function, then our results are congruent with a recent neuropsychological hypothesis that the female brain is less well lateralized—that is, manifests less hemispheric specialization—than the male brain for visuospatial functions. (De Lacoste-Utamsing & Holloway, 1982, p. 1432)

But, there is no evidence that size of callosum is related to the number of fibers or the number of fibers connecting the two sides is correlated in any way with lateralization or nonlateralization of hemispheric function, nor is there a logical, a priori basis for such a claim. But even if there *were* a known correlation between number of fibers and symmetry, still, no evidence whatsoever links the degree of symmetry or asymmetry with visuospatial ability. But the wording (quoted above) implies a link between symmetry and inferior function by the use of the terms *less well lateralized* and *less hemispheric specialization* (specialization being equated with superior ability) with respect to the female brain. If one were to rely on intuitive judgment in the interpretation of these inconclusive findings, one could just as easily suggest that a larger splenium indicates richer interhemispheric interchange, better integration of information and, therefore, increased bilateral representation of visuospatial perception. Thus, it could be predicted that women would be superior to men in visuospatial skills if they received the same training and experience (in three-dimensional perceptual/motor skills) during childhood as the majority of boys do. In short, it is equally possible bilateral or symmetical representation improves visuospatial perception. *No one knows.*

TESTOSTERONE, LEFT HANDEDNESS, AND MATHEMATICAL ABILITY

Another recent and well-publicized study (Geschwind & Behan, 1982) reported an association between left handedness, certain disorders of the immune system, and developmental learning disabilities such as autism, dyslexia, or stuttering, which are more common in boys. In attempting to explain this association of left handedness and, therefore presumably, right hemispheric dominance in boys, Geschwind and Behan quote a study of human fetal brains (Chi, Dooling, & Gilles, 1977) indicating that two convolutions of the right hemisphere develop 1 to 2 weeks earlier during gestation than those on the left. Geschwind and Behan proposed that testosterone (the sex hormone secreted by fetal testes but not by fetal ovaries) has the effect *in utero* of slowing the development of the left hemispheric cortex (the layers of neurons on the surface of the hemispheres). They offer this as an explanation for right hemispheric dominance in males. Yet, there is no evidence whatsoever for such an inhibitory effect of testosterone on the cortex in general nor do the authors acknowledge any scientific problem in their suggestion that circulating testosterone could selectively affect two convolutions of the left cortex alone.

Even more important to note is that Chi et al., in the same study quoted by Geschwind and Behan, did measurements on 507 brains from infants of 10–44-weeks' gestational age and stated clearly that they found *no significant sex differences* in their measurements. If testosterone had such an effect on the developing brain, there would *have* to be detectable sex differences in the timing of development of the convolutions of the left hemisphere in a series of this size.

Ignoring the findings by Chi et al. on *human* brains that clearly undermine their theory, Geschwind and Behan offered as supporting evidence for their hypothesis a study on *rats'* brains. That study (Diamond, Dowling, & Johnson, 1981) found that in male rats, two areas of the cortex that are believed to process visual information are 3% thicker on the right side than on the left; this difference was not found in female rats. The authors interpreted their findings to suggest that "in the male rat it is necessary to have greater spatial orientation to interact with a female rat during estrus and to integrate that input into a meaningful output" (p. 266). We see in this interpretation an unsupported conceptual leap from the finding of a thicker right cortex in the male to the assumption of "greater spatial orientation" in male rats. Yet there is no evidence that "spatial orientation" is related to any asymmetry of the cortex, nor that female rats have a "lesser" or somehow deficient spatial orientation.

One suspects that beliefs in men's superior visuospatial abilities and in their reported functional lateralization to the right hemisphere became hidden premises for the interpretation of the findings in this study on rats. The results of the study, including its unsupported conjectures, were then used by others as evidence for their own otherwise unsupported conjectures that testosterone enhances right hemispheric development and, therefore, superior mathematical ability in boys and men.

Whatever the significance of a thicker right cortex may be for the male rat's behaviors (a significance that is presently unknown), it surely is even more obscure what significance the rat's thicker right cortex has for *human* brains or behaviors. It is simplistic to suggest such an extrapolation from the rat's to the human cortex or behaviors in view of the enormous qualitative and quantitative leaps that the human brain and behavior have made in evolutionary development.

Though Geschwind and Behan did not study (or report) the incidence of giftedness in their population, it is predictable from the thesis of this paper that Geschwind would suggest in news reports of his work (in *Science*) that testosterone effects on the fetal brain can produce "superior right hemisphere talents, such as artistic, musical, or mathematical talent" (Kolata, 1983, p. 1312).

This proposal was greeted with enthusiasm by two investigators who had earlier reported a larger number of seventh-grade boys (260) scoring above 700 on the mathematics section of the Scholastics Aptitude Test (SAT) than seventh-grade girls (20), and had suggested an "endogenous" (i.e., innate), superior mathematical ability in boys (Benbow and Stanley, 1980, 1983). Criticisms of this work by mathematicians and other scientists have been numerous (Beckwith & Woodruff, 1984; Chipman, 1981; Fox, 1984; Schafer & Gray, 1981) and I shall not repeat them here, but it is important to acknowledge that no known measure, including the SAT, gauges "innate" intelligence or mathematical ability, nor is it clear how innate intelligence or mathematical ability *could* be measured or could even exist apart from learning and experience.

Since we do not know how or what brain structures and processes account for verbal fluency, mathematical skill, intelligence, or even consciousness, or for the enormous range of differences within any given population of people, such as any group of boys or girls, it is then not possible to explain *sex differences* in these processes. It is truly remarkable that such interest (to say nothing of time, talent, and money) is concentrated on the trivial sex differences that are found in some studies when systematic investigations of the sources of the great differences that are found *within* any same sex group could produce findings of potentially enormous social significance.

FABLE

It may, perhaps, be a welcome relief to read, instead of criticism, a fable to illustrate the problems with the studies of sex differences that I have discussed. In a North American Indian tribe, whom I shall call the Wisconsins, women have long been wondrous weavers, renowned for the exquisite color and patterned intricacy of their weaving, in a tradition passed through the generations from mothers to daughters or nieces. Since all women have always woven, there has been ample opportunity for many outstanding artists to emerge and be recognized. Thirty of them are now ranked among the world's top weavers. Over the past century, 45 Wisconsin women have become Nobel Prize winners in weaving. Currently, there are two outstanding Wisconsin weavers who are men and over the century, only two became Nobel laureates, one jointly with his wife, who is reputed to be the main source of their inspired and creative joint productions. Since weaving has always been a female tradition, men have entered the ranks of weaving with great difficulty. Their interest has generally been met with condescension and good-natured but firm discouragement of their efforts as being too clumsy and lacking in imagination, concentration, and seriousness. Therefore, few men have been formally trained and have had to pick up their skills by watching secretly and practicing at night when the weavers are asleep.

Now, my question is, in reading this fable would anyone seriously claim that there are so few outstanding male weavers because of their biological disabilities; that it is because (extrapolating from studies in the rat) their right visual and motor cortex is probably thicker (or more specialized) than the left; and, as is intuitively obvious because of the ambidexterity, computational skills, and artistic gestalt involved, superior weaving requires bilateral and equal cortical representation, as is the case in women? Furthermore, the claim could be made (by some obviously biased scientist) that this female superiority is determined by the effects of estrogen on the developing fetal brain. While all fetuses are exposed to high levels of maternal estrogens, it could be claimed that they affect only female fetuses since females have no testosterone to oppose the growth stimulating effects of estrogen on the cortex of *both* sides of the brain. This interpretation is at least as plausible as the testosterone story. In fact, weaving aside, this makes as reasonable a hypothesis to test for the question of sex differences in brain development and cognitive ability as the current ruling paradigm: Estrogen stimulates neuronal growth; testosterone may block estrogen effects; all fetal male brains may suffer from the effects of testosterone's inhibiting the growth stimulating action of maternal estrogens. Thus, the official exclusion of women from formal education and training in Western culture for the past 5,000 years (until the past 100 or so years), the cultural and institutional sanctions against physical training and achievement by women until recently, and the legal and physical restraints placed upon women's freedom and autonomy may have been the creative solution to a serious problem: that extraordinary cultural efforts were necessary to overcome (and obscure) a natural male handicap or deficiency, induced by testosterone.

The point I wish to emphasize again is not that this is a hypothesis that I seriously (very seriously) propose. Rather, I suggest that it has no less merit, logic, and imagination than the proposal of Geschwind and Behan, which was so readily accepted as valid by the interested scientific community.

SCIENTIFIC LANGUAGE AND THE RESPONSIBILITIES OF AUTHORSHIP

The body of work I have summarized has significance far beyond an ordinary controversy among academics. A combination of valid and irrelevant findings, flimsy evidence, and unsupported conjectures (strengthened by the omission of contradictory data) are assembled by the scientists and taken up by the national news services and major news media as demonstrating that male mathematical superiority may be from "math genes" or an "in-womb hormone" (The [Madison, Wisconsin] Capital Times, January 6, 1984, p. 3). The news item in *Science* reporting Geschwind's findings and his subsequent speculations carried the headline "Math Genius May Have Hormonal Basis" (Kolata, 1983). Such "scientific" claims for undemonstrated innate intellectual limitations on those who are not male can have only devastating effects on girls and women, except for those with rare and irrepressible talents.

While challenging and novel hypotheses are always welcome, it is expected that *some* supporting evidence exist for them, that explanations be offered for the more implausible assumptions implicit in an hypothesis (e.g., how circulating testosterone could have a unilateral effect on the cortex), and that knowledge of the existence of a major body of contradictory evidence not be suppressed. Most scientists have commitments to their hypotheses; such commitments may provide much of the energy and motivation to do research. But they serve science poorly if they distort the nature of the evidence to favor particular hypotheses and rely on popular belief for acceptance of hypotheses that lack the support of validating data.

At issue is the responsibility authors must assume for the variety of interpretations that can or inevitably will be made of their writing *because* of the ambiguities within, or the incompleteness or biasing of, the presentation. And the problems of understanding and interpretation of scientific reports are not limited to the nonscientific reading public. In this time of extreme specialization and technical sophistication, each of us who is a scientist must usually assume, on the basis of little more than faith, the carefulness and validity of the studies of other scientists, not only scientists in other fields but also those in other areas of our own field. Such validity rests not only on the design and execution of studies, but the completeness and candidness with which investigators discuss the uncertainties of their studies as well as the contradictory data that exist and the alternative interpretations that are possible. Few scientists have the knowledge of the techniques or database in other fields that would permit them to supply what the author omits. When, in addition to the usual problems of scientific understanding, the question under investigation has important social and political implications that are likely to be used to the detriment of one group or another, scientists cannot ignore their responsibility for the uses to which their writings may be put. While any scientific statement, however cautious, may be distorted by readers and science reporters, the final responsibility rests with scientists (along with science writers) who, intentionally or not, endow reports of their work with more certainty and relevance than the evidence warrants.

Scientists believe the language they use is simply a vehicle for the transmission of information about the objects of their research, another part of the scientist's toolbox, separate from the scientist's subjectivity, values, and beliefs. They do not recognize or acknowledge the degree to which their scientific writing itself participates in *producing* the reality they wish to present nor would scientists acknowledge the multiplicity of meanings of their text. The writers and readers of scientific texts must begin to pay heed to literary theorists. Contemporary literary textual criticism explodes the "comfortable relationship" between the word and what it represents, it disturbs the apparent "harmony of form and content" (Furman, 1980, p. 48). As literary criticism has "debunked the myth of linguistic neutrality" in the literary text, it is time to debunk the myth of the neutrality of the scientific text. The facade of neutrality and the techniques of scientific language (for example, the required passive voice) create the illusion of objectivity and anonymity, which, as the literary critic Terry Eagleton has pointed out, contributes to the authority of the text.

> The text does not allow the reader to see how the facts it contains were selected, what was excluded, why these facts were organized in this particular way, what assumptions governed this process, what forms of work went into the making of the text, and how all of this might have been different. Part of the power

of such texts thus lies in their suppression of what might be called their modes of production, how they got to be what they are . . . (Eagleton, 1983, p. 170)

But, in addition, as Furman has written, "Not only is the literary text a transmitter of explicit and implied cultural values, but the reader as well is a carrier of perceptual prejudices. It is the reader's acumen, expectations, and unconsciousness which invest the text with meaning" (p. 49). This is no less true of the scientific text. The place where the full meaning of the text is made, where the ambiguities are resolved is in the reader; that is where the multiplicities of interpretations, significances, and assumptions act and are resolved in some particular meaning. In Roland Barthes' words, "a text's unity lies not in its origin but in its destination" (1977, p. 148).

The meaning of the words of Geschwind and Diamond that I quoted, for example, is not fixed; it does not reside alone in the words as they lie on the page or as they represent authorial intentions or thought (conscious or unconscious), but in the *reading* of it by other scientists, by science writers, and the public who reads scientific reporting. Today that reading public is large and it, including its scientists, is in important respects a cultural unity. However unreflective the process may be, scientists, like those I quoted, are able to stop just short of making the kinds of assertions that their own and others' data cannot defensibly support, yet they can remain secure in the knowledge that their readers—other scientists and the very interested and susceptible public—will supply the relevant cultural meaning to their text; for example, that women *are* innately inferior in the visuospatial and, therefore, the mathematical skills, and that no amount of education or social change can abolish this biological gap. It is disingenuous for scientists to pretend ignorance of their readers' beliefs and expectations, and unethical to disclaim responsibility for the effects of their work and for presumed misinterpretations of their "pure" text. Scientists are responsible since they themselves build ambiguities and misinterpretations into the writing itself.

ALTERNATIVE HYPOTHESES

Underlying efforts to measure and define gender differences and to discover the biological forces shaping sexually differentiated behaviors and characteristics is the ancient dichotomy between biology and learning. The evidence that we have permits one important conceptualization of the development of cognition and behavior: biological and environmental factors are inextricable, in ways that make futile any efforts to separate them and measure *how much* of human behavior can be attributed to biology and *how much* to environment and learning. This is not to deny the great importance of efforts to understand the effects that individual factors, whether genes, hormones, neurons, or sensory input and learning, may have on development and behavior, but rather to emphasize the extraordinary complexity of the processes that define what is uniquely human: our mind, its creativity, and its nearly limitless capacity to learn.

At every stage of fetal development, genes, cells, the fetal organism as a whole, its maternal environment, and the external environment are in continuous interaction with each other, and each of these elements—including genes and their effects—continually change in response to these interactions.

Such processes of interaction and transformation are no less true for the brain, the organ of mind and behavior. We know that the critical events in the fetal development of the brain—the migration of nerve cells (neurons) from their genesis to their final positions, their survival, growth, and functional synaptic connections with other neurons—are not "simply" programmed genetically, since these basic patterns can be disrupted by deviations from the normal range of fetal environmental influences, as caused by viruses, chemicals, or metabolic abnormalities. Furthermore, the human brain is born relatively more immature than that of any other primate or other mammal. It doubles in size by the end of the first year of life and quadruples by the end of the fourth year. That means that the major growth of the human brain occurs precisely during the period of development in which it is exposed to a massive input of new sensory information from the external world. The major increase occurs in the size and complexity of the neurons and their dendritic and axonal ramifications and, therefore, in the extent and complexity of synaptic connections among neurons.

Furthermore, we know, from experimental work with animals and studies of the human brain, that the brain and neurons in sensory systems *require* sensory input for normal structural and functional development. Without input of light and visual stimuli to one or both eyes of cats and monkeys, neurons processing visual information fail to develop normal dendrites and normal connections (Weisel & Hubel, 1963). Similarly, the auditory system depends upon environmental stimulation for normal development. Auditory neurons remain structurally and functionally immature in mice reared with partial or complete sound deprivation, and profound structural changes in auditory neurons were found in a child born with sensorineural deafness (Trune, 1982; Webster & Webster, 1977, 1979). Thus the biology of the brain—the structure and functioning of its neurons—is itself molded by learning, as well as by genetic factors and other environmental influences.

It is because learning and environment are inextricable from the structure of neurons and because we have the property of mind, each mind the unique product of our individual, complex histories of development and experience, that I view as futile efforts to reduce human behaviors to biological parameters. Rather than biology, it is the cultures that our brains have created that most severely limit our visions and the potentialities for the fullest possible development of each individual.

REFERENCES

Alper, J. S. (1985). Sex differences in brain asymmetry: A critical analysis. *Feminist Studies, 11*, 7–37.

Barthes, R. (1977). *Image, music, text*. New York: Hill and Wang.

Beckwith, J., & Woodruff, M. (1984). Letter to the editor. *Science, 223*, 1247–1248.

Benbow, C., & Stanley, J. (1980). Sex differences in mathematical ability: Fact or artifact? *Science, 210*, 1262–1264.

Benbow, C., & Stanley, J. (1983). Sex differences in mathematical reasoning ability: More facts. *Science, 222*, 1029–1031.

Bleier, R. (1984). *Science and gender: A critique of biology and its theories on women*. Elmsford, NY: Pergamon.

Caplan, P. J., MacPherson, G. M., & Tobin, P. (1985). Do sex-related differences in spatial abilities exist? *American Psychologist, 40*, 786–799.

Chi, J. G., Dooling, E. C., & Gilles, F. H. (1977). Gyral development of the human brain. *Annals of Neurology, 1*, 86–93.

Chipman, S. (1981). Letter to the editor. *Science, 212*, 114–116.

Conner, R. L., & Levine, S. (1969). Hormonal influences on aggressive behaviour. In S. Garattini & E. B. Sigg (Eds.), *Aggressive behaviour*. Amsterdam: Excerpta Medica.

De Lacoste-Utamsing, C., & Holloway, R. L. (1982). Sexual dimorphism in the human corpus callosum. *Science, 216*, 1431–1432.

Diamond, M. C., Dowling, G. A., & Johnson, R. E. (1981). Morphologic cerebral cortical asymmetry in male and female rats. *Experimental Neurology, 71*, 261–268.

Eagleton, T. (1983). *Literary theory*. Minneapolis, MN: University of Minnesota Press.

Edwards, D. A. (1969). Early androgen stimulation and aggressive behavior in male and female mice. *Physiology and Behavior, 4*, 333–338.

Ehrhardt, A. & Baker, S. (1974). Fetal androgens, human central nervous system differentiation, and behavior sex differences. In R. C. Friedman, R. M. Richart, & R. L. Vande Wiele (Eds.), *Sex differences in behavior*. New York: Wiley & Sons.

Ehrhardt, A., Epstein, R., & Money, J. (1968). Fetal androgens and female gender identity in the early-treated adrenogenital syndrome. *Johns Hopkins Medical Journal, 122*, 160–167.

Ehrhardt, A., & Meyer-Bahlburg, H. (1979). Psychosexual development: an examination of the role of prenatal hormones. *Sex, Hormones and Behavior, 62*, 41–57.

Ehrhardt, A., & Money, J. (1967). Progestin-induced hermaphroditism: IQ and psychosexual identity in a study of ten girls. *Journal of Sex Research, 3*, 83–100.

Fairweather, H. (1976). Sex differences in cognition. *Cognition, 4*, 231–280.

Fee, E. (1979). Nineteenth century craniology: The study of the female skull. *Bulletin of the History of Medicine, 53*, 415–433.

Fox, L. H. (1984). Letter to the editor. *Science, 224*, 1292–1294.

Furman, M. (1980). Textual feminism. In S. M. Ginet, R. Borker, & N. Furman (Eds.), *Women and language in literature and society*. New York: Praeger.

Geschwind, N., & Behan, P. (1982). Left-handedness: Association with immune disease, migraine, and developmental learning disorder. *Proceedings of National Academy of Sciences, 79*, 5097–5100.

Gould, S. J. (1981). *The mismeasure of man*. New York: Norton.
Kessler, S., & McKenna, W. (1978). *Gender. An ethnomethodological approach*. New York: Wiley & Sons.
Kimball, M. M. (1981). Women and science: A critique of biological theories. *International Journal of Women's Studies, 4*, 318–338.
Kinsbourne, M. (1980). If sex differences in brain lateralization exist, they have yet to be discovered. *The Behavioral and Brain Sciences, 3*, 241–242.
Kolata, G. (1983). Math genius may have hormonal basis. *Science, 222*, 1312.
Lewontin, R. C., Rose, S., & Kamin, L. J. (1984). *Not in our genes. Biology, ideology, and human nature*. New York: Pantheon.
McGlone, J. (1980). Sex differences in human brain asymmetry: A critical survey. *The Behavioral and Brain Sciences, 3*, 215–263.
Mendelsohn, E., Weingart, P., & Whitley, R. (1977). *The social production of scientific knowledge*. Boston: Reidel.
Money, J., & Ehrhardt, A. (1972). *Man and woman, boy and girl*. Baltimore, MD: The Johns Hopkins University Press.
Provine, W. B. (1973). Geneticists and the biology of race crossing. *Science, 182*, 790–796.
Schafer, A., & Gray, M. (1981). Sex and mathematics. *Science, 211*, 229.
Trune, D. (1982). Influence of neonatal cochlear removal on the development of mouse cochlear nucleus. 1. Number, size and density of its neurons. *Journal of Comparative Neurology, 209*, 409–424.
Webster, D., & Webster, M. (1977). Neonatal sound deprivation affects brain stem auditory nuclei. *Archives of Otolaryngology, 103*, 392–396.
Webster, D., & Webster, M. (1979). Effects of neonatal conductive hearing loss on brain stem auditory nuclei. *Annals of Otolaryngology, 88*, 684–688.
Weisel, T., & Hubel, D. (1963). Effects of visual deprivation on morphology and physiology of cells in the cat's lateral geniculate body. *Journal of Neurophysiology, 26*, 978–993.
Westkott, M. (1984). Out of the shadows. *Women's Review of Books, 1*, 7–8.

Chapter 8

The Relationship Between Women's Studies and Women in Science

Sue V. Rosser

Women's studies has now been present in the academy for well over a decade. In her recent book, *Myths of Coeducation*, Florence Howe states:

> The seventies were not quiet for women engaged in developing 452 women's studies programs, 40 centers for research on women, 30,000 women's studies courses, more than 500 women's centers, hundreds of centers for reentry women, programs of continuing education for women, rape-crisis centers, committees on sexual harassment, committees on the status of women, not to mention the conferences, literary and arts festivals, special theatre and film programs, and musical celebrations. Thousands of new institutions and varieties of feminist programming now exist that were unknown only a decade ago, involving students, faculty, administrators, staff, and members of the community, mainly women, and women of color as well as white women. (1984, p. x)

The statistics collected by Betty Vetter (1981) indicate that more women have received advanced degrees in science and engineering in the 1970s and 1980s than in the 1950s and 1960s. In *Technology Review*, Hornig (1984) reported that nearly 4,500 women earn PhDs in science and engineering each year compared with only 1,500 in the humanities.

Large numbers of women, then, have been entering scientific fields at the same time as women's studies courses and programs have flourished. I propose to investigate the relationship between these two phenomena. Do these two groups represent interconnected parts of a diverse network that inform and support each other in pursuit of common goals by different means? Or in fact is there little, if any, relationship between the groups, who are taking relatively independent paths and are largely unaware of the connections in many of their parallel pursuits?

Much of the readily available, statistical information would certainly support a positive answer to the latter question. Most of the women's studies programs were founded as a direct outgrowth of the women's movement in the 1960s and 1970s. Certainly, women's health and reproductive rights have been two major issues of the movement in the 20th century. Health and reproduction in the view of U.S. 20th century medicine, are intimately connected to science, particularly to biology. One might have anticipated, therefore, that biology, health, and other branches of science might have been the first areas to be approached and developed in the academic arm of the women's movement known as women's studies. Yet, invariably on most campuses, the first women's studies courses were courses in literature or history. These were followed by courses in sociology, psychology, political science, philosophy, religion, and other disciplines in the social sciences and humanities. Courses in women's health and/or sexuality were usually the first science-related courses to appear in most women's studies programs. In many institutions, they appear well after the first wave of women's studies courses and often represent the only "science" courses that the program offers today. Few colleges and universities offer higher level courses in science such as Women's Studies 530: Science and Gender (University of Wisconsin, Madison); Biology 383: Advanced Topics in Biology, Women in Science: History, Careers, and Forces for Theoretical Change (Mary Baldwin College); or LBS 490E: Women, Nature, and Science/Technology (Michigan State University), which explore feminism and science at a theoretical level. A survey of a small nonrandom sample of colleges that offer certificates in women's studies (Rosser, 1986) revealed that more programs do not require a science or women's health course for the certificate in women's studies than do make such a demand.

There is also a dearth of scientists on or affiliated with the faculties of most women's studies programs. Many long-established programs (Towson State University; University of Maryland, College Park) have no faculty from the sciences or health as part of their programs. The paucity of scientists becomes evident at women's studies meetings if one simply counts the number of papers on the program from science or health-related fields. For example, at the National Women's Studies Association (NWSA) in 1984, 90 out of 908, or 10% of the papers on the program were on topics of science and/or health; 27 out of 268 or 10% of the panels or workshops dealt exclusively with science and/or health (Douglass College, 1984). Why aren't there more women scientists affiliated with women's studies?

One answer to this question is that there still are not very many women in science. Although Vetter's (1981) statistics indicate that more women have received advanced degrees in science and engineering in the 1970s and 1980s than in the 1950s and 1960s, the 1984 Report of the National Science Foundation documents that many fewer women (10.7%) than men (89.3%) receive degrees in sciences and engineering (exclusive of the social sciences where women comprise 34.6% of the population). The salaries, promotion, and advancement rates of women scientists are lower than men scientists at all ranks (this is not true for women in engineering) and unemployment rates for women are higher (NSF, 1984). Fewer papers concerning feminist issues and women's science and/or health issues appear on the program at national science meetings – 15 out of 184 papers at the American Association for the Advancement of Science (AAAS) in 1984, with none of the 19 panels devoted exclusively to such issues (AAAS, 1984) – than science and health papers appear on the program at women's studies meetings. The National Academy of Sciences, with a membership of 1,300, has only 29 or 30 women members, almost all of whom were elected within the past 6 or 7 years (Gornick, 1983). Thus, the women who are in science, particularly those with the higher ranking positions, still find themselves in the vast minority.

Because of their minority position, most women fear, and rightly so, that any commitments such as those to women's studies will make them appear more peripheral to traditional science and thus lessen their chances for promotion and tenure. The documented discrimination against women in science (Vetter, 1980) and the perception of many scientists that good science requires that all one's energy and thoughts be directed toward pursuing the problem in the laboratory substantiate their fears about the effects that involvement with anything outside of science (including women's studies) might have on their careers. Vivian Gornick quotes a successful woman scientist who expresses her fear that one of her women graduate students, who is getting married, will not have a life in science.

> "She's as interested in her upcoming wedding as she is in the experiments she's doing. I can't believe it." Alice shakes her head no. "She'll never make it. That's not what it's all about. At this point in her life she should have nothing – nothing on her mind but the lab. She should be *killing* herself with work. There should be absolutely nothing else in her head." (Gornick, 1983, p. 152)

Further, Gornick's work (1983) and that of E. F. Keller (1985) in particular have emphasized the extent to which 20th century science as it is pursued in the United States is synonymous with a masculine approach to the physical and natural world. Most women are unlikely to be attracted to the scientific approach as long as it is perceived as masculine (Kahle, 1983); those women who do become scientists often state that they must view the world from a masculine perspective in order to succeed professionally.

> She says, in essence, the only way to be a woman in science is to forget about being a woman. It is impossible to live in the world of contemporary professional science, and rise to the top of the profession, and still be a woman in old-fashioned terms (that is, have a family). She says it can't be done, and points out that the majority of women in science are unmarried, or married with no children, or divorced with no intention of remarrying. (Gornick, 1983, pp. 153–154)

The feminist perspective, which places women as the central focus for study, is thus directly antithetical to the scientific approach, if science is, in fact, synonymous with a masculine perspective. Only a few women in science who are very courageous and/or committed to feminism will be able to join women's studies programs and provide the link between women in science and women's studies.

During the last decade, a core of women scientists have attempted to make those links to women's studies. Women such as Bleier (1984), Hubbard (1979), and Keller (1985) fashioned critiques of traditional science and developed courses in women's studies. Historians of science (Fee, 1981; Haraway, 1978; and Hein, 1981) discussed some of the parameters and dualisms (subjectivity–objectivity; nature–nurture; dominant–subordinate) of the current androcentric science. The work of the feminists who are scientists and historians of science has led to an excellent critique of the current science and some hints of where changes might occur that might lead to a feminist science (Bleier, 1984). However, no feminist science has yet evolved. Although Fee (1982) stated that it may be impossible to develop a feminist science within the confines of a sexist society, the lack of a feminist science has led to discouragement on the part of many feminists in science and women's studies. Indeed, within the last eight months, I heard three feminists in science, each associated for more than a decade with a feminist critique of science, deplore the fact that we have not been able to develop in science the sorts of far-reaching theoretical changes that feminist scholars evolved for most disciplines in the social sciences (Leavitt, 1975; Leibowitz, 1975) and humanities (Daly, 1978). They question whether feminists in science will ever move beyond demonstrating the unscientific biases behind the current biologically deterministic theories such as sociobiology, where genes are assumed to determine behavior, and endocrinology, where hormone levels are assumed to differentially affect brain and behavior in males and females, to develop new paradigms and theories that change the traditional androcentric theories of current science. Two of the three are so discouraged that they stated publicly that they have stopped work on a feminist science and returned to full-time work on their traditional scientific research.

The statements of these feminists in science troubled me deeply. They led me to explore the steps and phases taken by feminists in other disciplines to derive the theoretical schemes that have transformed studies in the humanities and social sciences. Very little has been written about the process that leads to theoretical transformation. Virtually all of the work in this area (McIntosh, 1983; Arch and Kirschner, 1984; Minnich, 1982; Howe, 1984; Tetreault, 1985; Schuster and Van Dyne, 1984) focuses primarily upon curricular change and transformation. However, I think that it might be useful to appropriate one of these schemes and apply it to the research, teaching, and personal development of women in science. Although the schemes were not intended for this usage, I thought this approach might bring out some of the relationships between women's studies and women in science and might shed some light on why a theoretical transformation to a feminist science had not yet occurred.

Schuster and Van Dyne (1984) developed the scheme in Table 8.1 to assess curriculum transformation.

The teaching of science in most institutions has been affected very little by the feminist transformation. Most courses outside of women's studies programs are probably taught at the first stage, where the absence of women is not noted. The "factual" nature of most science courses leads instructors to emphasize information derived from experiments. Few professors consider for themselves, and fewer still attempt to convey to the students, the parameters of social, historical, and gender bias that may influence the theories derived from interpretation of the facts. Many professors think that gender is not a bias that influences hypotheses, subjects, experimental design, or theory formation in science; therefore the absence of women is not noted.

The image of the professor as authority and the student as "vessel" is perhaps a more dominant model of pedagogy in the sciences than in the social sciences and humanities (Rosser, 1986). The German model for science education, the complex and expensive laboratory instrumentation, and the highly technical language and jargon of science further reinforce the professor's authority over knowledge in the classroom and laboratory. Furthermore, many science courses serve as preprofessional training for students seeking entrance to medical, veterinary, or dental school or for students who will be licensed by state boards in nursing, medical technology, or radiology. The push for "maintaining 'standards of excellence'," which frequently exclude women, comes not only from colleagues within the science faculty but from links to the professional scientific hierarchy outside the college or university. Thus, a variety of pressures and practices, both inside and outside academic science, reinforce professors who fail to notice the absence of women in their teaching.

Some professors have noted the absence of women in science. Although a few simply state that the absence of great women in science proves the inferiority of women or women scientists' failure to do significant research (Cole, 1979), most are eager to progress to the next stage, to find the women who have made it in the male scientific world and add them on at the appropriate place on the syllabus. Certainly Marie Curie gets added to most such courses. Other women who have won Nobel Prizes (Irene Joliot-Curie, Gerty Theresa Cori, Maria Goeppert Mayer, Dorothy Crowfoot Hodgkin, Rosalyn S. Yalow, and Barbara McClintock) (Haber, 1979) are likely to be mentioned. A simple, but effective technique that helps to bring to light more women scientists who have made equal contributions to those of men, is to mention the first, as well as last, names of experimenters (for example, Dr. Herbert M. Evans and Dr. Gladys Anderson Emerson, when discussing the

Table 8.1. Assessing Curriculum Transformation

Stages	Questions	Incentives	Means	Outcome
Absence of women not noted	Who are the truly great thinkers/actors in history?	Maintaining "standards of excellence"	Back to basics	Pre-1960s exclusionary core curriculum Student as "vessel"
Search for missing women	Who are the great women —the female Shakespeares, Napoleons, Darwins?	Affirmative action/ Compensatory	Add on to existing data within conventional paradigms	"Exceptional" women on male syllabus Student's needs recognized
Women as disadvantaged, subordinate group	Why are there so few women leaders? Why are women's roles devalued?	Anger/Social justice	Protest existing paradigms but within perspective of dominant group	"Images of women" courses "Women in politics" Women's Studies begins Links with ethnic, cross-cultural studies

Women studied on own terms	What was/is women's experience? What are differences among women? (attention to race, class, cultural difference)	Intellectual	Outside existing paradigms; develop insider's perspective	Women-focused courses Interdisciplinary courses Student values own experience
Women as challenge to disciplines	How valid are current definitions of historical periods, greatness, norms for behavior? How must our questions change to account for women's experience, diversity, difference?	Epistemology	Testing the paradigms Gender as category of analysis	Beginnings of integration Theory courses Student collaborates in learning
Transformed, "balanced" curriculum	How can women's and men's experience be understood together? How do class and race intersect with gender?	Inclusive vision of human experience based on difference, diversity, not sameness, generalization	Transform the paradigms	Reconceptualized, inclusive core Transformed introductory courses Empowering of student

Source: From "Placing Women in the Liberal Arts: Stages of Curriculum Transformation," by M. Schuster and S. Van Dyne, November 1984, *Harvard Educational Review, 54*, p. 419. Copyright 1984 by President and Fellows of Harvard College. Used with permission.

isolation of vitamin E from wheat germ oil and the study of its function). If first names are not mentioned, the stereotype running through most students' heads will be that of the white male scientist. Professors who add women to their syllabi usually are gratified by the positive response of the women students who, particularly in classes for science majors, are desperately seeking role models. The professor is gratified by having reached the women students in a more effective way.

After searching more carefully for the "great women of science," a few professors begin to feel uncomfortable and may move on to the third stage, with its inevitable question of why are there so few women scientists. Particularly if the professor is a woman, she may become very angry and begin to analyze the forces that have kept women in the lower ranks or out of science entirely. She may respond to or seek overtures with the women's studies program or faculty in other disciplines seeking ways to include women in their teaching. In traditional science courses, she may ask students to read and discuss *Rosalind Franklin and DNA* (Sayre, 1975) along with *The Double Helix* (Watson, 1968) to illustrate the treatment of women in science. Recognizing that not only women, but essentially anyone who is not a white man has been excluded from science, may lead her to ask students to read *Black Apollo of Science* (Manning, 1983), about the life of E. E. Just.

Professors who are teaching traditional science courses at this stage often bring in additional statistical information about the dearth of women in science (National Academy of Sciences, 1979; National Science Foundation, 1984) and some social factors (Martin & Irvine, 1982) that may be contributing to that dearth. Special programs for women with math anxiety, or special attention to classroom techniques (pairing women with women, rather than men with women lab partners) may be used to aid women students seeking science careers in the traditionally androcentric science programs.

Other professors teaching at this stage may develop courses such as Women in the History of Science (history of science faculty) or Women and the Health Care System (health sciences faculty), which accept the traditional approaches to the study of history of science and health care but ask how women fit into these paradigms. Most of the women's studies courses or health courses that focus on women's bodies are also taught within the traditional paradigms of this stage.

A few women's studies courses focus on the relationship between women and science. These courses and some courses centered on women's bodies and health fall into the fourth stage. They place women in the center of the analysis and use an interdisciplinary (Biology and Psychology of Women, Science and Gender) approach to the study of problems. Primarily due to a strong push from the women's self-health movement outside the academy, professors in these courses have begun to question the paradigms of their traditional disciplines and of the health care system. Women's actual experience, as compared to the physician's analysis or scientific theory, is valued in these courses. However, not a great deal of attention has been paid as yet to the varieties of experience and differences among women due to race, class, sexual preference, and cultural diversity. For example, during the summer of 1984, the first conference on black women's health care was held in Atlanta (Billye Avery, the Director and Founder of the National Black Women's Health Project organized the conference). In sum, I suggest that in terms of teaching, scientists teaching women's studies are primarily at the third stage, with a few teaching courses at the fourth stage. Most women scientists not teaching women's studies courses are probably teaching within the first two stages.

The Schuster–Van Dyne scheme developed to assess stages of curriculum transformation can also be applied to research. Research provides the data, materials for the theories, and the perspectives that form the basis of material taught in the classroom. At what stage is the research done by women scientists and/or feminists practicing science?

In applying the Schuster–Van Dyne scheme to research, one must recognize that the "outcomes" would differ since they would not be directly applicable to courses. As in the case of teaching, most men researchers will be operating at the first stage. The problem of researchers failing to notice the absence of women is exacerbated in the sciences by the supposed unbiased and objective scientific method. Other feminists (Bleier, 1984; Fee, 1982; Hein, 1981; Keller, 1985) and I (1984) have written about the androcentric bias in choice of questions asked, subjects, experimental designs, interpretations of data, and theories drawn from data. However, many scientists contend that science is gender-free and that the scientific method permits no bias (Keller, 1985), including that of gender. Unfortunately, some women in science are very resistant to recognition of the androcentric bias. Gornick (1983) writes of the women who had to accept the male perspective of the scientific world to survive in it.

Some researchers, both men and women, have begun to look for the missing women in scientific research. By this, I do not mean the search for women scientists of the past whose accomplishments have been ignored, misunderstood, or credited to someone else; this is a valuable, necessary search that the historians of science (Fee, 1982; Rossiter, 1982) are now undertaking. I am referring directly to scientific research, originally carried out on males of a species, which is now undertaken using the females of that species. This research adds to the existing data within conventional paradigms. Two examples, one from endocrinology and one from primatology, illustrate this stage-two research. In much endocrine research, unless hormones of the reproductive system were being studied, male rats were used as experimental subjects. The noncyclicity of the males was thought to provide a "cleaner" baseline to examine nonreproductive hormone fluctuations; male rats are often also cheaper to purchase. Now, many of the experiments originally run solely on males are being repeated using virtually the same protocols on females.

In primatological research, many women primatologists originally went out in the field to examine the behavior of the female primates of various species. Initially, the women primatologists were not feminists (Hrdy, this volume) and they accepted the findings of the earlier men primatologists regarding primate behavior. However, they recognized that little data had been collected on aspects of female primate behavior, other than mother–infant interactions. Their aim, then, was to gather this data.

In both endocrinology and primatology, the search for the missing females in scientific experiments often led directly into third- and fourth-stage research. For example, Hoffman's work (1982) with female rats showed that the cyclicity of the female rat might more accurately depict the complex realities of episodic hormone release than that of the noncyclic male. This then led researchers to question other variables such as age, species, and individual variation, which challenge the current theory of hormone fluctuations (fourth stage).

Sarah Blaffer Hrdy describes, in this volume, the progression of her own field work in which she began to recognize that the theories she had learned in graduate school for primate behavior did not apply to the female langurs she was observing (third stage). The transition from third-stage to fourth-stage research was concurrent with a rise in feminist consciousness on the part of the researcher, who also began to seek affiliation with other women's studies scholars at this time. In her chapter in this volume Hrdy states

> In my own case, changes in the way I looked at female langurs were linked to a dawning awareness of male-female power relationships in my own life . . . and only towards the end of the 1970s had I begun to read anything by feminist scholars like Carolyn Heilbrun and Jean Baker Miller. . . . That is, there were two (possibly more) interconnected processes: an identification with other females among monkeys taking place at roughly the same time as a change in my definition of women and my ability to identify and articulate the problems women confront. (Hrdy, 1986, p. 140)

She then began to question exactly what was the experience of female primates and what were the variations among species of primates (fourth stage). Hrdy (this volume) has now reached what I would consider fifth-stage research. She definitely uses gender as a category of analysis. By changing the questions asked to include competition among women, female agency in sexuality, and infanticide, she is testing the traditional paradigms. She states that her own gender is crucial for her interest in female primate behavior and for preparing her to question the existing paradigms of animal behavior. "The notion of 'solidarity' with other women and, indeed, the possibility that female primates generally might confront shared problems was beginning to stir and to raise explicit questions about male–female relations in the animals I studied" (Hrdy, 1986, p. 140). Hrdy still considers the basic paradigm of sociobiology to be acceptable. But, unlike the most prominent men sociobiologists, she regards sociobiology as a body of theories and interpretations that require constant testing and are likely to undergo change to allow the data observed to fit the theory for females with regard to competition, sexual selection, and infanticide.

Primatology is the field within the sciences where the research has been most transformed by the feminist perspective. Most subdisciplines, even within biology, remain at the first stage. Increasingly, however, researchers are including females as subjects in their experiments (second stage), conducting research on subjects of primary interest to women, such as menopause, childbirth, and menstruation/estrus (third stage), and recognizing the limits to generalizing beyond the data collected on limited samples to other genders, species, and conditions not sampled in the experimental protocol (fourth stage) (Bleier, 1984). This will undoubtedly lead to fifth-stage research in subdisciplines other than primatology.

The personal development of women in science may be the most difficult to assess using the Schuster–Van Dyne scheme. The interviews of contemporary women scientists reported by Vivian Gornick in *Women in Science: Portraits from a World in Transition* (1983) provide a sample of women scientists who vary in their knowledge and acceptance of feminism and in their opinions regarding gender and science. I might add that my own observations of women colleagues in science substantiate the reports of Gornick's work.

Although, in fact, very few women are in science, most of the men and some women scientists appear not to have noticed their absence. Some of these women assumed that they were entering a masculine world when they became scientists; they made the choice to "act like men" and took it for granted that there would be few other women who would make that choice.

> A woman scientist of only thirty years ago was a lady eccentric, a denatured bluestocking, a nineteenth-century New Woman. She wore tweeds and oxfords, cared nothing for love, kept a cat, and smoked cigarettes. Adopting the style of the gentlemen scientists among whom they worked, these women acceded in a socially repressive atmosphere to an even deeper repression of self. (Gornick, 1983, p. 120)

Other women at the first stage have not only accepted the gender-free and objective nature of science on the theoretical level, they also assume that women are absent from science because other women aren't good enough to be scientists. This "queen bee syndrome" is a particular version of a more general story: a member of a subordinate group attempts to identify with the oppressor.

Fortunately, most women scientists have reached at least the second stage in their personal development; they recognize that very few women are in science. They begin to question the old-boy network and request affirmative action hiring in departments where women with PhDs are hired as research associates rather than as professors in tenure-track positions as the men are.

> In 1960 chemistry department heads said as openly as they had in 1940: "We don't hire women." In 1980, of course, no one would dare say that. Yet, at one of the great research universities one of the Mrs. Godblesses told me: "The chemistry department here doesn't advertise. It's illegal now, but they still do it that way. Somehow, they consider it a 'shame' to advertise. They write to their friends. And of course their friends are men who have only male graduate students. But even so, some awfully good young women get through the system and come up here for interviews. It's always the same. They look at these excellent young women and they say, 'She's very good but she lacks seasoning. Let her go off somewhere else for the year and then we'll consider her again. Of the young men just like her they say, 'We'd better grab him before someone else does.'" (Gornick, 1983, pp. 102–103)

At this stage, women scientists may organize on a university- or companywide level to gather statistics on the numbers of women at different levels in the organization.

The information gleaned from these statistics quickly leads many women scientists to the third stage and to feminism on a personal level. Knowledge of their poor position and the discrimination against women in science leads to anger. The recognition of the scientific hierarchy, in which the top-level, decision-making positions are held by men and the lower-level, technician positions, where the "dirty work" is done, are held by women often leads to demands for social justice. Some women in science become so angry and frustrated at this stage that they actually leave science.

> One of them, Louise Anderson, was a research associate for her husband for sixteen years. She thought hers was the perfect working life. Then another associate, a woman much older than herself, said to her, "There'll come a time, you'll want to do certain things, and you won't be able to, and then you'll start going backward, there's no way to stop it," and suddenly Louise realized how restless she was. From there to resentful was one short step. She left Yale, her husband, and science within a year. (Gornick, 1983, p. 79)

Fortunately, most choose not to leave science but to explore the ways by which other women have coped within the current system of science (third stage). At this stage, women scientists are cognizant of the androcentric shortcomings of the system and search for ways whereby they can manipulate the system for themselves and other women. They search for women (or men who really want to help women rise in the system) as role models and mentors. They develop formal and informal "old-girl" networks and support groups that will help women at all levels of the scientific hierarchy (Briscoe, 1984). Many will seek contacts with women's studies programs or colleagues in other disciplines who are feminists. A great many of the women's caucuses within professional societies are functioning at this stage, trying to improve the lot of women within the existing power structure (Briscoe, 1984).

> "After that Leon and I decided we were going to get me a job or go to court. And that's what finally happened. We got these jobs here, with tenure for him when we came, and the firm promise of tenure for me in five years. When the five years were up they denied me tenure, and we sued. It was such a long fight. And it wasn't the physicists who got me tenure, it was the feminists. As I said, I'm tired now." (Gornick, 1983, p. 89)

Some of the women within these groups begin, after battling the system to increase chances for women for a long period of time, to question the underlying philosophy of an androcentric science and system, fourth stage concerns. They wonder whether a science based on a masculine perspective of the world and demanding characteristics defined in our society as masculine, such as objectivity, rationality, and dominance, not only excludes women but also the best interests of humanity in general. Perhaps scientists who recognize that the theories of science develop in a context of social and historical values feel some responsibility for the potential benefits and risks of the applications of those theories. Maybe a shortening of the distance between the subject and the observer so that the scientist has some feeling or empathy with the organism is desirable (Keller, 1983). Possibly a more holistic, contextual approach to problems would cause less harm to our environment than the mechanistic, reductionistic, hierarchical model of much of modern science (Merchant, 1979). At the fifth stage, women scientists recognize that the stereotypical feminine characteristics, such as subjectivity, feeling, and dependence, may be the very attributes that need to be brought to science to prevent the annihilation of all of humanity. Women scientists at this stage realize that gender must play a crucial role in science. Women have a unique, central contribution to make toward changing the paradigms of the current androcentric science.

A few women scientists, according to Gornick's interviews, have made the transition to the sixth stage. They recognize that there is more than one way to practice science. For some people the traditional "all day and night, every day in the lab" may be best. For others, a totally different time schedule and approach may lead to other valuable insights.

> Look at it this way. My chairman is a brilliant fellow, but he's not really interested in science anymore. He's interested in travel, theater, I don't know *what* the hell he's interested in, half the time he's not here at all. But when he *is* here, he's invariably doing interesting and potentially important work. That man operates at ten percent of his capacity. He's working on a ninety-percent margin. Now the guy, down the hall, he's a competent scientist. Not brillant at all. Just hardworking and industrious. He's in the lab ninety hours a week and, believe me, he needs every one of those ninety hours to accomplish the little that he does accomplish. He's working on a ten-percent margin. Me, I'm somewhere between my chairman and the guy down the hall. I have to be here every day, and put in a good eight hours in the lab, but I don't have to be here ninety hours a week to produce good work. (Gornick, 1983, p. 156)

These women realize that diversity in race, class, gender and work style among scientists who have different ways of approaching problems and practicing science will be necessary to bring about the transformations so needed in current scientific paradigms. They have made the transition in their personal development to the sixth stage, which is very likely the necessary prerequisite for transformation of existing research paradigms. Women scientists at this stage of personal development may be on the threshold of breakthroughs to sixth-stage research, which might also, in its turn, generate changes in teaching.

In reviewing the application of the Schuster–Van Dyne scheme to the work of women scientists, a pattern emerges. Although most women scientists may be at the first or second stage in teaching, research, and personal development, some have reached later stages. It is in the area of personal development that the highest stage has been reached, according to this analysis. In research, no one has gone beyond the fifth stage, while in teaching maximum development seems to be at the fourth stage. This progression suggests that women in science are touched by the women's movement and by women's studies first in their personal lives. Only when this occurs do women scientists have new insights that begin to transform our research. Teaching is the most difficult area to change. It is affected last by new perceptions, since it is through research that reconceptualization takes place. Thus it seems that while women's studies and women in science are not usually interconnected in a direct way, the impact of women's studies is being strongly felt by many women in science. On this foundation, further relationships between the two groups can be built, relationships that will strengthen and enrich both.

REFERENCES

AAAS Annual Meeting Preconvention Program. (1984, March 30). *Science, 23*, 1381–1392.

Arch, E., & Kirschner, S. (1984). Transformation of the curriculum: Problems of conception and deception. *Women's Studies International Forum, 7*, no. 3, 149–151.

Bleier, R. (1984). *Science and gender: A critique of biology and its theories on women.* Elmsford, NY: Pergamon Press.

Briscoe, A. M. (1984). Scientific sexism: The world of chemistry. In V. B. Haas and C. C. Perrucci (Eds.), *Women in scientific and engineering professions.* Ann Arbor: The University of Michigan Press.

Cole, J. R. (1979). *Fair science: Women in the scientific community.* New York: Free Press.

Daly, M. (1978). *GynEcology: The metaethics of radical feminism.* Boston: Beacon Press.

Douglass College. (1984). *Steering our course: Feminist education in the '80's.* NWSA Conference Program, June 24–28.

Fee, E. (1981). Is feminism a threat to scientific objectivity? *International Journal of Women's Studies, 4*, no. 4, 213–233.

Fee, E. (1982). A feminist critique of scientific objectivity. *Science for the People, 14*, no. 4, 8.

Gornick, V. (1983). *Women in science: Portraits from a world in transition*. New York: Simon and Schuster.
Haber, L. (1979). *Women pioneers of science*. New York: Harcourt Brace Jovanovich.
Haraway, D. (1978). Animal sociology and a natural economy of the body politic, Part I: A political physiology of dominance; and Animal sociology and a natural economy of the body politic, Part II: The past is the contested zone: Human nature and theories of production and reproduction in primate behavior studies. *Signs, 4*, no. 1, 21–60.
Hein, H. (1981). Women and science: Fitting men to think about nature. *International Journal of Women's Studies, 4*, 369–377.
Hoffman, J. C. (1982). Biorhythms in human reproduction: The not-so-steady states. *Signs, 7*, no. 4, 829–844.
Hornig, L. (1984, Dec.). Women in technology. *Technology Review*.
Howe, F. (1984). *Myths of coeducation*. Bloomington: Indiana University Press.
Hrdy, S. (1986). Empathy, Polyandry and the Myth of the Coy Female. In R. Bleier (Ed.), *Feminist perspectives on science*. Elmsford, NY: Pergamon Press.
Hubbard, R. (1979). Have only men evolved? In R. Hubbard, M. S. Henifin, and B. Fried (Eds.), *Women look at biology looking at women*. Cambridge, MA: Schenkman.
Kahle, J. (1983). The disadvantaged majority: Science education for women. Burlington, NC: Carolina Biological Supply Company. AETS Outstanding Paper for 1983.
Keller, E. (1983). *A feeling for the organism: The life and work of Barbara McClintock*. New York: W. H. Freeman.
Keller, E. (1985). *Reflections on gender and science*. New Haven, CT: Yale University Press.
Leavitt, R. R. (1975). *Peaceable primates and gentle people: Anthropological approaches to women's studies*. New York: Harper and Row.
Leibowitz, L. (1975). Perspectives in the evolution of sex differences. In R. R. Reiter (Ed.), *Toward an anthropology of women*. New York: Monthly Review Press.
Manning, K. (1983). *Black Apollo of science: The life of Ernest Everett Just*. New York: Oxford University Press.
Martin, B. R., & Irvine, J. (1982). Women in science – The astronomical brain drain. *Women's Studies International Forum, 5*, no. 1, 41–68.
McIntosh, P. (1983). Interactive phases of curricular re-vision: A feminist perspective. Working Paper No. 124. Wellesley, MA: Wellesley College, Center for Research on Women.
Merchant, C. (1979). *The death of nature: Women, ecology and the scientific revolution*. New York: Harper and Row.
Minnich, E. (1983). Friends and critics: The feminist academy. In C. Bunch and S. Pollack, *Learning our way: Essays in feminist education*, pp. 317–330. Trumansburg, NY: The Crossing Press.
National Academy of Sciences. (1979). *Climbing the academic ladder: Doctoral women scientists in academe*. Report of the Committee on the Education and Employment of Women in Science and Engineering. Washington, DC: Commission on Human Resources.
National Science Foundation. (1984). *Women and minorities in science and engineering*. Report 84-300.
Rosser, S. V. (1984). Call for a feminist science. *International Journal of Women's Studies, 7*, no. 1, 3–9.
Rosser, S. V. (1986). *Teaching about science and health from a feminist perspective:*

A practical guide. Elmsford, NY: Pergamon Press.

Rossiter, M. W. (1982). *Women scientists in America: Struggles and strategies to 1940*. Baltimore, MD: The Johns Hopkins University Press.

Sayre, A. (1975). *Rosalind Franklin and DNA: A vivid view of what it is like to be a gifted woman in an especially male profession*. New York: W. W. Norton & Company, Inc.

Schuster, M., & Van Dyne, S. (1984). Placing women in the liberal arts: Stages of curriculum transformation. *Harvard Educational Review, 54*, no. 4, 413–428.

Tetreault, M. K. (1985, July). Stages of thinking about women: An experience-derived evaluation model. *The Journal of Higher Education*.

Vetter, B. (1980, March). Sex discrimination in the halls of science. *Chemical and Engineering News*, 37–38.

Vetter, B. (1981). Degree completion by women and minorities in science increases. *Science, 212*, no. 3.

Watson, J. D. (1968). *The double helix*. New York: Atheneum Publishers.

Chapter 9

Taking Feminist Science to the Classroom: Where Do We Go From Here?

Mariamne H. Whatley

If the theoretical work represented by the papers in this volume is to have an impact beyond reinforcing the views of already committed feminists, both scientists and nonscientists, it must reach the classroom in accessible form. The critiques of science and biological determinism and the visions for a feminist science can become more than academic rhetoric by translation into lay language to influence the consumers of both the products and the media byproducts of scientific research. The vision often presented is that the current university science curriculum would be transformed (Rosser, 1986), and eventually this transformation would filter down, perhaps via school of education classes, to the high schools. However, it is not enough to wait for this transformation to take place at the college level. For this analysis to have any effect in terms of perceptions of science, committed teachers at all levels must begin to revise their approaches to teaching the sciences and any other subjects that draw on scientific research.

The purpose of this chapter is to discuss the crucial educational function of countering the unquestioning acceptance by many nonscientists of any scientific or pseudoscientific information presented. My approach will be divided into two sections: challenging the search for biological explanations for human behavior and questioning the belief in experimentalism. Specific examples will be discussed, which can be used in teaching in a wide range of classrooms. In other papers on feminist revisions of science education, the emphasis is often on the need to attract more women into science, thereby helping transform science by numbers alone (Kelly, White, & Smail, 1984; Rose, 1984; Rosser, 1986). This might include such tactics, discussed by Rosser in this volume, as calling attention to women scientists who have made the kinds of contributions to science that are publicly recognized (i.e., winning the Nobel Prize) or examining reasons why more women have not entered into science, particularly its higher echelons. While this is a very important role education can play in bringing feminist concerns into science, more women in science is not the answer if women continue to be trained to practice science the same way men have been trained in the past. Through education, we must change both the way science is practiced and the way the public responds to it.

The scientific method is highly valued in our society, especially since the effort to challenge Sputnik began a race for scientific dominance and placed a premium on scientific training. Many seem to believe that if enough money goes into research, any problem can be solved: curing cancer, finding a renewable safe energy source, developing the sure deterrent to nuclear war. Accompanying the faith that science can eventually solve all problems is the belief that it can also provide answers and explanations. For example, when *Sociobiology* (Wilson, 1975) was published with dramatic publicity, its premises, including extrapolation from animal to human behavior, were readily accepted by the general public. Sociobiologists began explaining behaviors as diverse as nonmonogamy, altruism, and war in terms of our genes and our evolutionary history. Faith in science partially explains why biological causes are eagerly accepted as explanations for complex human behavior. But this acceptance goes beyond mere trust that science can provide answers, for these explanations serve to remove the burden of choice and responsibility in certain areas, both on a narrow personal level and a broader political level, shifting these burdens to biological entities. The danger is that we no longer correctly view genes and hormones as discrete biochemical structures; instead, they are seen as very abstract powerful forces controlling our lives.

First, looking at the level of personal choice, it is easy to describe specific examples of ways we minimize personal responsibility by ascribing causality to uncontrollable biological forces. The following examples could be used effectively in health-related courses to approach this issue. While certain diseases, such as breast cancer and cardiovascular disease, may have a hereditary component, it is important to put those factors into perspective with other risks. On the positive side, knowledge that a family history increases risk of heart attack can motivate an individual to stop smoking, reduce cholesterol intake, and begin a program of aerobic exercise. However, by placing the burden on heredity, two problems are possible. One is that some people may be very fatalistic when, actually, much can be done to reduce the risk. On the other hand, those whose risk is low in terms of family history may not perceive themselves as needing to reduce other known risks. It is easy to forget that when the term *family history* is used, this refers both to hereditary and environmental components. If people continue high fat, high cholesterol eating patterns learned at home as children, this part of the family history may be much more devastating than the genetic component in terms of heart disease.

Another related example is the impact of having a family history of breast cancer, which is generally described as increasing risk two to three times. This can easily mean that women may fail to examine other risk factors that can be controlled, such as high animal fat consumption. Playing on women's fears of breast cancer by emphasizing the hereditary component can lead to medical abuses such as prophylactic mastectomy. This surgical removal of healthy breast tissues, which is then replaced with implants, is often suggested for "high risk" women. How *risk* is defined, however, is a problem, especially since a high percentage of women with breast cancer had no identifiable risk factors. The concept of family history as a risk factor must be carefully examined. One woman who had a prophylactic mastectomy did so because her mother and aunt both had breast cancer. What was not considered in the assessment of risk that led to the surgery was that both her mother and aunt had tuberculosis as adolescents and received almost daily x rays for treatment and diagnostic purposes. This radiation greatly increased their risk of breast cancer but in no way affected the relatives' chances. The belief that breast cancer is genetically inevitable or highly likely can lead to unnecessary surgery (surgery that also ironically makes it harder to identify cancer if it does arise in the remaining tissue), to ignoring ways other risks may be reduced, and possibly to an unfounded sense of safety for those who perceive their risk as low.

Another example of the reduction of a complex issue to simple biological factors involves the controversy surrounding the description of premenstrual syndrome (PMS) as a hormone deficiency disease. This is an excellent topic for health, psychology, or sexuality classes, since it illustrates both the eagerness with which biological explanations are accepted and the difficulty disentangling the strands of hormonal influence and cultural expectations. Certainly, the recognition that the very small percentage of women who suffer severe debilitating symptoms on a monthly basis are not crazy but, in fact, suffer from real physiological reactions was an important breakthrough in validating women's own experiences and definitions of health and illness. However, the overextension of this theory to explain all depression, anxiety, and panic attacks can easily bring us back to an argument, for example, for a biological basis for the small numbers of women in high pressure political positions, as well as clearly providing a rationale for not putting any energy or money into correcting this situation. This reliance on hormonal factors may also cause us to ignore the fact that our premenstrual mood changes, while possibly exaggerated by hormonal influences, probably have a real basis in the problems in our lives. A male colleague and I decided that our similar anxiety attacks probably had a lot more to do with pretenure syndrome than any hormonal condition. Anxiety may also correlate more closely to a monthly cycle of low bank balances than of hormonal fluctuation.

The emphasis on hormonal causes also totally ignores the point that the few women who are most "disabled" by PMS have the advantage of being able to predict the times they will encounter difficulties and, therefore, can develop coping strategies, whereas men in power may suffer *unpredictable* mood swings. Also, chronic conditions that may affect a man's work all the time are not seen as prohibitive in the way that part-time problems with PMS are viewed. Focusing on the biological issues distracts us from looking at socioeconomic issues and the question of who benefits from the definition of PMS as a serious disease.

PMS provides a new way of dismissing the validity of women's problems and complaints on one level, while validating them on another; it can be used to trivialize women's concerns and reinforce stereotypes. A statement by the husband of one of the leading figues in PMS research, in the Foreword to one of her books, illustrates the latter approach.

> The old cliche, "It's a woman's privilege to change her mind," calls for an even greater tolerance than before now that it is realized that every woman is at the mercy of the constantly recurring ebb and flow of her hormones. (Dalton, 1969, p. viii)

It is interesting to note that while many texts on sexuality are quick to pick up on hormonal explanations for PMS, they often ignore the physiological cause of menstrual cramps (dysmenorrhea), substances called *prostaglandins* (Whatley, 1985). Biological explanations of PMS reinforce stereotypes of women's emotional instability that can then serve as a barrier to achievement, whereas the recognition of the physiological basis for dysmenorrhea, which can easily be treated with inhibitors of prostaglandin synthesis, means eliminating a barrier.

Another example of the way hormonal explanations have often figured in discussions of behavior is that of the role of testosterone in controlling libido, which texts on sexuality often emphasize, while leaving out crucial psychological and social factors. The assumption that testosterone level is a major factor in sex drive is then inappropriately used as an explanation for a wide variety of sexual behaviors, such as the double standard of adolescent sexual behavior or the prevalence of rape in our society. Continuing this illogical line of reasoning based on a false premise, elimination of unwanted or illegal sexual behaviors is seen as an issue of hormonal control. Recently, Depo-Provera, a hormone that has an antitestosterone effect, has been introduced for "chemical castration" as an alternative to prison for sex offenders. This example shows the ready acceptance of biological explanations as well as a denial of complex social issues such as the role of women in our society and the basis of the patriarchal power structure. Rape becomes incorrectly and dangerously reduced to an issue of sex drive and not one of power, dominance, and control. If we can reduce all complex social and cultural issues to simple biological explanations, we can cure everything and keep the drug companies in business. A discussion of the implication of using Depo-Provera to treat sex offenders is an excellent opportunity to counter biological determinist theories in sexuality, psychology, or law classes.

A discussion of the dependence of libido on biological factors has surfaced recently in another form. An article on sexuality noted that current research suggests that heredity has a great deal to do with libido (Penney, 1985, p. 53). While the article did discuss other factors involved in libido, the biological argument can make it easy for us to avoid examining either the different values and emphases we all put on the role of sexual activity in our lives or changes in libido in varied contexts, such as in specific relationships or at different stages in our lives. Instead we may accept an unexamined biological explanation, incorrectly assuming that if a woman is not interested in sexual activity the fault is in her genes.

Clearly, there are broad political implications in biological explanations of such issues as PMS and rape. In recent years, a number of authors, many of whom are scientists, have been examining ways in which biological theories have been and are still used to maintain certain inequalities in society (Bleier, 1984; Lowe & Hubbard, 1983; Sayers, 1982). Rather than going over the same ground, it is enough for me to mention that biological determinism has been used both to explain inequalities and to reinforce them; these inequalities include, but are not limited to, the right to vote, access to education, entry into different kinds of work and careers. In the 19th century, research on cranial capacity was carried out to attempt to prove inherent inferiority in intelligence of women compared to men and of blacks compared to whites. Today we laugh at those studies, while current research on race and IQ and on sex differences in brain lateralization correlated with differences in cognitive abilities carry out a similar approach (see Bleier, this volume). The misuse of biological theories becomes a powerful tool in the hands of the dominant group (whether that group is male, white, upper class, heterosexual) to prevent changes in laws or redistribution of resources that might help eliminate any of these inequities in the system. Just as people may willingly accept the artificial biological limitations on personal responsibility for personal behavior, they may easily accept limitations on personal and public responsibility for social problems. Opponents of affirmative action plans can happily latch on to any scientific theory that suggests that women may be biologically unqualified for certain jobs, either for physical reasons such as being too weak, or cognitive reasons such as lack of math ability. Whether women are blocked from construction work or overrepresented in minute piecework, someone can find a biological justification. Hormones, whether affecting the muscles or the brain, are invoked in such arguments.

How does a teacher fight against this willingness to accept biological determinism? Having taught women's health and biology for a number of years, I have been very aware of the initial eagerness students often show in trying to find biological explanations for many phenomena, sometimes trying to prove inherent biological superiority of women. Using the kinds of examples discussed in this paper, critiquing biological explanations of behavior and developing sociocultural theories are approaches that generate both scepticism of biological determinist arguments and much more creativity in looking for alternative and more complex and complete explanations. The students who enroll in women's studies classes come with great diversity in levels of consciousness of feminist issues and many are already very open to counter arguments. It is interesting to note, however, that some feminist scholars not only are wary of biological determinism but also translate that wariness into an avoidance of studying biology at all. Members of women's studies programs have mentioned that biology is not part of their women's studies curriculum because they view a discussion of women's biology as an acceptance of biological determinism. In other words, some women's studies scholars themselves have not worked through these issues and, therefore, are unable to deal with any biological determinist argument, feeling themselves on shaky ground. To ignore the biological arguments in teaching women's studies is to create more vulnerability to these theories and to give more power to those who use these arguments. These issues must be dealt with openly in all women's studies classes or the biological arguments will continue to carry too much weight.

In a reactionary time, the educational system as a whole seems geared towards acceptance of biological determinism. The backlash against feminism means emphasis on theories such as the biological determination of gender roles. For example, current texts being used to teach sexuality and health in universities often identify differences in male and female behavior as being hormonally controlled (Whatley, 1985), clearly an argument for maintaining gender-based social structures in our society. Students have reported unchallenged comments made by faculty members in a number of different classes such as "Homosexuality is known to be caused by prenatal effects of hormones on the brain" or "Since women don't produce any androgens, they cannot develop much muscle strength." Both statements are scientifically inaccurate but were reported as facts in lecture, noticed by women's studies students who were beginning to question scientific authority. In these cases, one important role of the politically committed teacher is to help provide information to correct these inaccuracies and another is to generate discussion on the implications of these issues. For example, it should be made clear in response to the latter comment that both the ovaries and adrenal cortex produce androgens in women and, furthermore, that adjusting for size, men and women are basically equal in lower body strength. What this implies in terms of physical capacity to do work traditionally considered too hard for women can then be discussed.

What is more crucial than just supplying more accurate information to use as ammunition in debates is to help students develop alternative hypotheses, to see the roles social, cultural, and political factors can play in what appear at first to be biological issues. This can be done easily outside of the women's studies classroom, including biology, health, physical education, political science, home economics, nursing, and medical classes. What appear to be biological issues can always be put in a broader context so that the interactions of biology and society can be examined. Taking the example of androgens and strength in women mentioned above, it is possible, in a physical education class, to examine any biological limitations of women in terms of physical abilities. After realizing the minimal role hormones and other biological factors actually do play, the class can look at nonbiological changes that occur around puberty that might serve to limit the development of women in terms of physical potential. In addition, it is possible to ask a series of questions about a given issue, after sifting through different explanations: What would be the implications (in terms of policy, etc.) of biological causation? What would be the implications of sociocultural causation? Of the interaction of these? Who might benefit by the acceptance of different explanations? Who might lose out? In the example of the physical education class, differences in allocation of resources for competitive sports and recreation can be examined as obvious outcomes.

What seems most important in the classroom to defuse the arguments of biological determinism is to create a basic distrust of simplistic scientific explanations for complex phenomena. Classroom discussions of why a biological explanation might be easier, rather than more likely to be accurate, is useful. This questioning in the classroom of why an explanation becomes accepted or popular can help students analyze media coverage of science. When an article in *Science* came out suggesting that males are inherently better in mathematics (Benbow & Stanley, 1980), it immediately received headlines suggesting the existence of a math gene. It is important for students to learn to view such a headline very sceptically and critically.

It is also important as part of the educational process to examine the scientific method and to call into question the faith in experimentalism that seems engrained in our society. The way science is commonly taught is to present an idealized view similar to this:

1. Scientist develops a hypothesis.
2. Scientist designs a perfect, definitive experiment to test the hypothesis.
3. Scientist carries out the experiment, which works the first time and forever after.
4. The hypothesis is proven and is now fact.
5. The fact is now presented in textbooks.

It can be a shock to anyone taught science in this way to encounter real science. For example, it is rarely discussed that a number of alternative hypotheses can often be supported by any one experiment, that experiments are often much easier to conceptualize than to carry out, that results are usually not clear cut. (Some geneticists believe that even Gregor Mendel had to fudge his data to get it to fit his hypothesis cleanly.) One way to approach this in class is to discuss experiments that did not work, in the sense that they weren't conclusive. Students can easily learn to examine real data and develop alternative hypotheses to explain the results.* Assignments to critique research papers are useful in a number of different courses so that students can get an idea of sampling techniques, appropriate controls, how to use statistics.

Since this approach still assumes a great deal about scientific truth, one way it can backfire is to create the belief that perfect experiments can be designed in any area, even if it involves highly complex questions. Part of the problem is that science teachers don't usually raise the issue that even "good" science reflects the values of the scientist. One woman scientist told me that the only problem she saw with the field of sociobiology is that some of the more highly popularized work is done by bad scientists. She continued to say that if science is done right, no political issues are involved. That common view ignores the issue that the very question asked in a research study can have political implications. It is essential to remind students that science, no matter how well it is done, can have political implications and that scientists have to take responsibility for what is done with their work.

One approach, which can be used in a wide range of classes, is to examine what research is actually being done in a particular field, what questions are being asked, and what is being ignored. A clear example to use in a women's studies class is the emphasis in research on gender differences rather than gender similarities. The choice of looking for differences assumes that they are probably there *and* that they are significant, making it possible to design an experiment likely to find these differences, and often resulting in the publication of papers that find such differences. Other examples that can generate a lot of discussion in different classes range from the paucity of research in the area of men's contraceptives to the low level of research on alternative energy resources. It is an easy step from these for students to begin examining who controls research and research funding and to destroy the image of neutral science. Creative teachers can find examples in any field: a home economics class might examine the scientific research going into labor-saving devices and why these devices have actually increased rather than decreased time spent in housework.

Having established in class that even research using the best scientific methodology is still limited by the values of both those supporting and those doing the research, teachers should again reinforce the point that equally

---

**Editor's note*: A classic paper worth reading and rereading by scientists and non-scientists, students and teachers, is T. C. Chamberlin's, The method of multiple working hypotheses, *Science* (old series) *15*, p. 92 (1890), reprinted in *Science, 148*, pp. 754–759 (1965).

careful scrutiny must be given to results that support a feminist viewpoint. There are usually feminist cheers accompanying every study, for example, that shows female superiority or no gender-based differences in a certain area. However, it is only a temporary measure to use scientific arguments to defuse antifeminist biological determinist theories; we must move beyond that to questioning the relevance of scientific data to issues of oppression based on gender, race, or class. For example, gender differences in verbal ability, visual–spatial ability, and mathematical ability have become accepted in a great deal of the literature on cognitive abilities, so that scientific data countering this would be very useful. Data from these studies were examined using the techniques of meta-analysis with the results that these differences, though "well-established" in the literature, were found to be very small (Hyde, 1981). Certainly this finding of minimal differences is very useful in providing data for arguing with antifeminist biological determinists. However, citing this research means acceptance of the whole body of research that is looking for gender differences. One question to ask ourselves is, If Hyde's results had shown much larger differences than expected, would we still accept the end product? There has to be some consistency about what we accept and don't accept and this must be based on more than our politics. It is crucial to consider whether this kind of research, whichever side it supports, should have any place in discussions of this highly charged political area.

Trying to challenge faith in biological determinism and the scientific method is a difficult but necessary task for a teacher. Few subjects are free of political implications, as has been illustrated by using examples appropriate to a number of different topics. Teaching without examining the status quo can imply acceptance of the status quo; someone who is committed to feminism can find ways in almost any subject at any level to raise feminist issues, no matter how they are labeled (or left unlabeled). Students outside of women's studies at the university level, as well as students at much earlier stages in their education, are very receptive to feminist arguments raised within the context of their areas of interest. Theoretical issues discussed in the preceding chapters in this book must be translated into concrete examples accessible to very diverse groups and it is the job of committed teachers to do this.

REFERENCES

Benbow, C. P., & Stanley, J. C. (1980). Sex differences in mathematical ability: Fact or artifact? *Science, 210*, 1262–1264.

Bleier, R. (1984). *Science and gender: A critique of biology and its theories on women*. Elmsford, NY: Pergamon.

Dalton, K. (1969). *The menstrual cycle*. New York: Random House.

Hyde, J. S. (1981). How large are cognitive gender differences? A meta-analysis using W^2 and d. *American Psychologist, 36*, 892–901.

Kelly, A., White, J., & Smail, B. (1984). *Girls into science and technology: Final*

report. Manchester, England: Manchester Polytechnic (mimeo).
Lowe, M., & Hubbard, R. (Eds.). (1983). *Woman's nature: Rationalizations of inequality*. Elmsford, NY: Pergamon.
Penney, A. (1985). Is it possible to have (good) sex with the same person for 20 years? *Ms., 13*, no. 12, 52–53, 94.
Rose, H. (1984). Nothing less than half the labs. In J. Finch and M. Rustin (Eds.), *Agenda for higher education*. Hammondsworth, England: Penguin.
Rosser, S. V. (1986). *Teaching science and health from a feminist perspective: A practical guide*. Elmsford, NY: Pergamon.
Sayers, J. (1982). *Biological politics: Feminist and anti-feminist perspectives*. New York: Tavistock Publications.
Whatley, M. H. (1985). Male and female hormones: Misinterpretations of biology in school health and sex education. In V. Sapiro (Ed.), *Biology and women's policy*. Beverly Hills, CA: Sage Publications.
Wilson, E. O. (1975). *Sociobiology: The new synthesis*. Cambridge, MA: Harvard University Press.

Chapter 10

Further Readings on Feminism and Science

Susan E. Searing

This bibliography highlights recent feminist critiques of scientific theory and practice. The literature in this emerging field is scant, but the listing is nonetheless selective. All cited items are in the English language. Books and contributions to books are emphasized, along with key journal articles. A few representative biographies and works about the lives and status of women scientists are included; occupational guidance materials are not. The compiler made a thorough search of the women's studies literature through June 1985. Significant new writings disseminated through the traditional channels of science and social science publishing may have been missed.

Aldrich, M. L. (1978, Autumn). "Review essay: Women in science." *Signs, 4*, 126–135.

Alic, M. (1984). *Hypatia's heritage: The history of women's science*. London: Pandora Press.

Arditti, R., Brennan, P., & Cavrak, S. (Eds.). (1980). *Science and liberation*. Boston: South End Press.

Arditti, R., Duelli-Klein, R., & Minden, S. (Eds.). (1984). *Test-tube women: What future for motherhood*? London: Pandora Press.

Arnold, L. B. (1984). *Four lives in science: Women's education in the nineteenth century*. New York: Schocken Books. (Biographies of Maria Martin Bachman, Almira Hart Lincoln Phelps, Louisa C. Allen Gregory, and Florence Bascom.)

Associations and committees of or for women in science, engineering, mathematics, and medicine. (1984, May). Washington, DC: Office of Opportunities in Science, American Association for the Advancement of Science.

Baldwin, R. S. (1981). *The fungus fighters: Two women scientists and their discovery*. Ithaca, NY: Cornell University Press. (Biographies of Elizabeth Hazen and Rachel Brown.)

Berryman, S. E. (1983). *Who will do science*? New York: Rockefeller Foundation.

Biology as destiny: Scientific fact or social bias? (1984). Cambridge, MA: Science for the People.

Birke, L., & Silverton, J. (Eds.). (1984). *More than the parts: Biology and politics*. London: Pluto Press.

Bleier, R. (1978). Social and political bias in science: An examination of animal studies and their generalizations to human behavior and evolution. In E. Tobach & B. Rosoff (Eds.), *Genes and Gender II*, pp. 49–69. New York: Gordian Press.

Bleier, R. (1984). *Science and gender: A critique of biology and its theories on women*. Elmsford, NY: Pergamon.

Brighton Women and Science Group. (1980). *Alice through the microscope: The power of science over women's lives*. London: Virago.

Briscoe, A. M., & Pfafflin, S. M. (1979). Expanding the role of women in the sciences. *Annals of the New York Academy of Science, 323*.

Chinn, P. Z. (1980). *Women in science and mathematics: Bibliography*. Arcata, CA: Humboldt State University.

Committee on the Education and Employment of Women in Science and Engineering, Office of Scientific and Engineering Personnel, National Research Council. (1983). *Climbing the ladder: An update on the status of doctoral women scientists and engineers*. Washington, DC: National Academy Press.

Dagg, A. I. (1983). *Harem and other horrors: Sexual bias in behavioural biology*. Waterloo, Ontario: Otter Press.

Easlea, B. (1981). *Science and sexual oppression: Patriarchy's confrontation with woman and nature*. London: Weidenfeld & Nicolson.

Easlea, B. (1983). *Fathering the unthinkable: Masculinity, scientists and the nuclear arms race*. London: Pluto Press.

Faulkner, W., & Arnold, E. (Eds.). (1985). *Smothered by invention: Technology in women's lives*. London: Pluto Press.

Fausto-Sterling, A. (1981). Women and Science. *Women's Studies International Quarterly, 4*, 41–50.

Fausto-Sterling, A. (1985). *Myths of gender: Biological theories about women and men*. New York: Basic Books.

Fee, E. (1981, Sept./Oct.). Is feminism a threat to scientific objectivity? *International Journal of Women's Studies, 4*, 378–392.

Fee, E. (1982, July/Aug.). A feminist critique of scientific objectivity. *Science for the People, 14*, 5–8, 3–33.

Fee, E. (1983). Women's nature and scientific objectivity. In M. Lowe & R. Hubbard (Eds.), *Woman's nature: Rationalizations of inequality*, pp. 9–28. Elmsford, NY: Pergamon.

Fox, L. H., et al. (1980). *Women and the mathematical mystique*. Baltimore, MD: Johns Hopkins University Press.

Goddard, N., & Henifin, M. S. (1984, Winter). A feminist approach to the biology of women. *Women's Studies Quarterly, 12*, 11–18.

Goodfield, J. (1981). *An imagined world: A story of scientific discovery*. New York: Harper & Row.

Goodman, M. J., & Goodman, L. E. (1981, Sept./Oct.). Is there a feminist biology? *International Journal of Women's Studies, 4*, 393–413.

Gornick, V. (1983). *Women in science: Portraits from a world in transition*. New York: Simon & Schuster.

Haas, V. B., & Perrucci, C. C. (Eds.). (1984). *Women in scientific and engineering professions*. Ann Arbor: University of Michigan Press.

Haraway, D. (1978, Autumn). Animal sociology and a natural economy of the body politic, Part I: A political physiology of dominance. *Signs,4*, 21–36.

Haraway, D. (1978, Autumn). Animal sociology and a natural economy of the body politic, Part II: The past is the contested zone: Human nature and theories of production and reproduction in primate behavior studies. *Signs, 4*, 37–60.

Haraway, D. (1981, Spring). In the beginning was the word: The genesis of biological theory. *Signs, 6*, 469–482.

Harding, S., & Hintikka, M. B. (Eds.). (1983). *Discovering reality: Feminist perspectives on epistemology, metaphysics, methodology, and philosophy of science*. Dordrecht, The Netherlands: Reidel.

Hein, H. (1981, Sept./Oct.). Women and science: Fitting men to think about nature. *International Journal of Women's Studies, 4*, 369–377.

Henifin, M. S. & Amatniek, J. C. (1982). Bibliography: Women, science, and health. In R. Hubbard, M. S. Henifin, & B. Fried (Eds.), *Biological woman – The convenient myth*, pp. 289–376. Cambridge, MA: Schenkman.

Hrdy, S. B. (1981). *The woman that never evolved.* Cambridge, MA: Harvard University Press.

Hubbard, R., (1981). The emperor doesn't wear any clothes: The impact of feminism on biology. In D. Spender (Ed.), *Men's studies modified*, pp. 213–235. Elmsford, NY: Pergamon.

Hubbard, R., Henifin, M. S., & Fried, B. (Eds.). (1979). *Women look at biology looking at women*. Cambridge, MA: Schenkman.

Hubbard, R., Henifin, M. S., & Fried, B. (Eds.). (1982). *Biological woman – The convenient myth*. Cambridge, MA: Schenkman.

Hubbard, R., & Lowe, M. (Eds.). (1978). *Genes and gender II: Pitfalls in research on sex and gender*. New York: Gordian Press.

Humphreys, S. (Ed.) (1982). *Women and minorities in science: Strategies for increasing participation*. Boulder, CO: Westview. (AAAS Selected Symposium No. 66.)

Keller, E. F. (1980). Feminist critique of science: A forward or backward move? *Fundamenta Scientiae, 1*, 341–349.

Keller, E. F. (1982). Feminism and Science. *Signs, 7*, 589–602. (Reprinted in N. O. Keohane, M. Z. Rosaldo, & B. C. Gelpi (Eds.), *Feminist Theory: A Critique of Ideology*. Chicago: University of Chicago Press, 1982.)

Keller, E. F. (1983). *A feeling for the organism: The life and work of Barbara McClintock*. San Francisco: W. H. Freeman.

Keller, E. F. (1983, Sept./Oct.). Feminism as an analytic tool for the study of science. *Academe, 69*, 15–21.

Keller, E. F. (1985). *Reflections on science and gender*. New Haven, CT: Yale University Press.

Kimball, M. M. (1981, Sept./Oct.). Women and science: A critique of biological theories. *International Journal of Women's Studies, 4*, 318–338.

Lewontin, R. C., Rose, S., & Kamin, L. J. (1984). *Not in our genes: Biology, ideology, and human nature*. New York: Pantheon.

Longino, H. E. (1981, Fall). Scientific objectivity and feminist theorizing. *Liberal Education, 67*, 187–195. (Reprinted in M. Triplette (Ed.), *Women's studies and the curriculum*, pp. 33–41. Winston-Salem, NC: Salem College, 1983.)

Longino, H., & Doell, R. (1983, Winter). Body, bias, and behavior: A comparative analysis of reasoning in two areas of biological science. *Signs, 9*, 206–227.

Lowe, M. (1983). The dialectic of biology and culture. In M. Lowe & R. Hubbard (Eds.), *Woman's nature: Rationalizations of inequality*, pp. 39–62. Elmsford, NY: Pergamon.

Malcolm, S. M., Hall, P. Q., & Brown, J. W. (1976). *The double bind: The price of being a minority woman in science*. Washington, DC: American Association for the Advancement of Science. (AAAS Report No. 76-R-3.).

Merchant, C. (1980). *The death of nature: Women, ecology and the scientific revolution*. New York: Harper & Row.

Messing, K. (1983). The scientific mystique: Can a white lab coat guarantee purity in the search for knowledge about the nature of women? In M. Lowe & R. Hubbard (Eds.), *Woman's nature: Rationalizations of inequality*, pp. 75–88. Elmsford, NY: Pergamon.

National Science Foundation. (1982 to present, biennial). *Women and minorities in science and engineering*. Washington, DC: National Science Foundation.

Newman, L. M. (Ed.). (1984). *Men's ideas/women's realities: Popular science, 1870–1915*. Elmsford, NY: Pergamon.

Osen, L. M. (1974). *Women in mathematics*. Cambridge, MA: M.I.T. Press.

Overfield, K. (1981). Dirty fingers, grime, and slag heaps: Purity and the scientific ethic. In D. Spender (Ed.), *Men's studies modified*, pp. 237–248. Elmsford, NY: Pergamon.

Ramaley, J. A. (Ed.). (1978). *Covert discrimination and women in the sciences*. Boulder, CO: Westview. (AAAS Selected Symposium No. 14.)

Reed, E. (1978). *Sexism and science*. New York: Pathfinder Press.

Richer, D. (Ed.). (1982). *Women scientists: The road to liberation*. London: Macmillan.

Rose, H. (1983, Autumn). Hand, brain, and heart: A feminist epistemology for the natural sciences. *Signs, 9*, 73–90.

Rosser, S. V. (1982). Androgyny and sociobiology. *International Journal of Women's Studies, 5*, 435–444.

Rosser, S. V. (1984, Jan./Feb.). A call for feminist science. *International Journal of Women's Studies, 7*, 3–9.

Rosser, S. V. (1985). Integrating the feminist perspective into courses in introductory biology. In M. R. Schuster & S. R. Van Dyne (Eds.), *Women's place in the academy: Transforming the liberal arts curriculum*, pp. 258–276. Totowa, NJ: Rowman & Allanheld.

Rossiter, M. W. (1982). *Women scientists in America: Struggles and strategies to 1940*. Baltimore, MD: Johns Hopkins University Press.

Rothschild, J. (Ed.). (1983). *Machina ex Dea: Feminist perspectives on technology*. Elmsford, NY: Pergamon.

Ruse, M. (1981). *Is science sexist? And other problems in the biomedical sciences*. Dordrecht, The Netherlands: Reidel.

Sayers, J. (1982). *Biological politics: Feminist and anti-feminist perspectives*. New York: Methuen.

Sayre, A. (1975). *Rosalind Franklin and DNA: A vivid view of what it is like to be a gifted woman in an especially male profession*. New York: Norton.

Science and technology (thematic issue). (1984, Summer). *Canadian Woman Studies/les Cahiers de la Femme, 5*.

Science, sex and society. (1979). Washington, DC: Women's Educational Equity Act Program, U.S. Dept. of Health, Education and Welfare.

Shaw, E., & Darling, J. (1984). *Female strategies*. New York: Walker.

Siegel, P. J., & Finley, K. T. (1985). *Women in the scientific search: An american bio-bibliography, 1724–1979*. Metuchen, NJ: Scarecrow Press.

Sloane, E. (1980). *Biology of women*. New York: Wiley & Sons.

Standish, L. (1982, Sept./Oct.). Women, work, and the scientific enterprise. *Science for the People, 14*, 12–19.

Stehelin, L. (1976). Science, women and ideology. In H. Rose & S. Rose (Eds.), *The radicalisation of science: Ideology of/in the natural sciences*, pp. 76–89. London: Macmillan.

Tobach, E., & Rosoff, B. (Eds.). (1978). *Genes and gender*. New York: Gordian Press.

Warner, D. J. (1979). *Graceanna Lewis, scientist and humanitarian*. Washington, DC: Smithsonian Institution Press.
Watts, M. W. (1984). *Biopolitics and Gender*. New York: Haworth Press. (Also published as *Women & Politics, 3*, no. 2/3, Summer/Fall, 1983.)
Women and science (thematic issue). (1981, Sept./Oct.). *International Journal of Women's Studies, 4*.
Women in science: A man's world (thematic issue). (1975). *Impact of Science on Society, 25*.
Women, science, and society (thematic issue). (1978). *Signs, 4*.
Women, technology and innovation (thematic issue). (1981). *Women's Studies International Quarterly, 4*.
Yee, C. Z. (1977, Fall). Do women in science and technology need the women's movement? *Frontiers, 2*, 125–128.
Zimmerman, J. (Ed.). (1983). *The technological woman: Interfacing with tomorrow*. New York: Praeger.

Index

About the Editor and Contributors

ABOUT THE EDITOR

Ruth Bleier

So far as I can tell now, I grew up not knowing girls and women were supposed to be inferior. Girls were always, with a few exceptions, the best students in my classes and my very intelligent mother had very intelligent women friends. And then, as an only child, my parents put all their eggs in this basket, and their assumption was always clear, if only implied, that I would or could do any or everything well: math, cooking, grammar, baseball, typing.

After deciding that my mother's plan for me to go out and cure the world of wars and injustice through a career in the diplomatic corps was unrealistic, I switched from political science to premed in my junior year at Goucher College then went to the Woman's Medical College of Pennsylvania, then the only remaining women's medical school in the country. There, I became involved in left radical politics, which met considerable official attention and opposition throughout the dismal period of McCarthy and the House Unamerican Activities Committee, but it was a way of life and thought for me that has persisted through many transformations.

I practiced general medicine in Baltimore for several years, then took postdoctoral training in neuroanatomy at the Johns Hopkins School of Medicine, continuing my research there until coming to the University of Wisconsin in Madison in 1967. In 1970, the women's movement refocused my activism and for the first time integrated my political and personal lives. I was a feminist campus activist and organizer, and became a member of the Women's Studies Program when it was established in 1975, in large measure a result of the active women's movement on campus.

In the early 1970s, I began to see how the biological sciences were affected by sexist and other cultural biases, and have been working since then to apply feminist analyses and perspectives to the theories and practices of science. My book in this area, *Science and Gender, A Critique of Biology and Its Theories on Women*, was published by Pergamon Press in 1984. In the other half of my professional life, where I am a professor in the Department of

Neurophysiology, I use light and electron microscopes to study the brain. I am interested both in the organization of the hypothalamus as it relates to function and in the roles of a population of scavenger cells that reside within the third ventricle, a central space in the brain filled with cerebrospinal fluid.

THE CONTRIBUTORS

Elizabeth Fee

I was born in Belfast, Northern Ireland, and, as the child of Methodist missionary parents, spent much of my early life in China and Malaysia. My adolescence was spent in the midst of the political struggles of Northern Ireland in the early 1960s. I began to explore mathematics in the hope of discovering a form of truth independent of religion or politics but, on being told that mathematics was unsuitable for a girl, obediently turned to the biological sciences. I studied biochemistry at Cambridge University until I became fascinated by the history and philosophy of science and especially by the idea of the relativity of scientific knowledge. In 1968, I came to Princeton University, where I studied the history of science. Much of my time at graduate school in the late 1960s and early 1970s was spent discovering the women's movement and radical politics. Prompted by feminist friends, I wrote my doctoral dissertation on science and feminism: a history of scientific theories about women and sex differences in Victorian Britain.

I spent 2 years at the State University of New York at Binghamton, teaching courses on sexuality, feminist theory, and the history of science, then several years at the School of Health Services at Johns Hopkins University, where I taught courses on the history, politics, and future of medicine. When the school closed after a succession of political and financial crises, I moved to the School of Hygiene and Public Health, an older and more established branch of Johns Hopkins, but one that still preserves some of the more progressive traditions of public health.

At the School of Public Health, I have been making myself a new identity as a historian of public health, concerned with the ways in which the disciplines of public health have historically been shaped and created, the construction of knowledge about health and disease, and the politics of public health programs nationally and internationally. I am currently completing a book on the history of public health education. I have also continued my earlier interest in the history of science and gender, and have published a variety of articles on the history of anthropology, psychology, evolutionary theory, and feminist epistemology. I edited *Women and Health: The Politics of Sex in Medicine* and am currently working with a group of graduate students and faculty at Johns Hopkins on a collection of original articles on the politics of women's health. In this area, I am especially interested in reproductive rights and the social and political implications of the new reproductive technologies.

Donna Haraway

For many years my principal concerns have centered on feminism, science, and technology, in relation both to the gendered social relations of science and to systems of ideology and cultural meanings. I am presently completing a book on the multiple contestations in late capitalist society for scientific and popular meanings of race, sex, and class in the bodies of monkeys and apes. The book will be called *Primate Visions: A History of the Craft of Story Telling in the Sciences of Monkeys and Apes*. My other writing is about "cyborgs," a science fiction image of machine-animal-human hybrids that has become prominent in women's science fiction writing in the last decade, as well as an image that hints at the complexities of women's symbolic and material positions in "high technology" worlds. With graduate students at the University of California at Santa Cruz, I am writing about feminist science fiction as political theory—and vice versa. I am currently a professor, teaching women's studies, feminist theory, and the history of science in the History of Consciousness Department at the University of California at Santa Cruz. History of Consciousness is an interdisciplinary doctoral program in the humanities at UCSC; within the program about 30 people are working toward PhDs with a central emphasis in feminist theory.

I have been part of feminist women's movements since about 1970 and helped found the Women's Studies Program at the University of Hawaii in the early 1970s. I earned my PhD at Yale in Biology, but actually worked in an interdepartmental arrangement in history of science, philosophy, and biology for the last 2 years of graduate school. I taught at the University of Hawaii from 1970–1974, then moved to the History of Science Department of Johns Hopkins from 1974–1980, where I taught courses about biology and politics. During those years, I worked in the Women's Union in Baltimore and participated in feminist theory discussion groups with many fine women, whose work has helped me immensely. I was also part of Baltimore's Science for the People. In 1976, I published *Crystals, Fabrics, and Fields: Metaphors of Organicism in 20th Century Developmental Biology* (Yale University Press). Most of my other writing from that period addressed aspects of the history and politics of biology within feminist theory. In 1980, I moved to UC Santa Cruz, to a tenured position in History of Consciousness in feminist theory made possible by the concerted work of feminists there over several years. At UCSC, I was coordinator of the Women's Studies Program for 2 years and am a founding member of the Silicon Valley Research Project, whose major concerns include the politics of gender, race, and class in the technologically mediated, new international division of labor. I have also worked to develop curriculum and broad campus concern about the social studies of science and technology from feminist perspectives.

Sarah Blaffer Hrdy

As a primatologist who is also a feminist (and by *feminist* I mean someone who advocates equal rights and opportunities for both sexes; the term takes on political dimensions only when—as is often the case—countervailing sentiments or conditions deny women equal consideration), I have attempted to present evidence about females and to analyze behavior from female as well as male points of view. Traditional explanatory models for primate social organization cast females in passive or primarily nurturing roles, and the result was distorted and often quite mistaken models of primate breeding systems. Hence, the last decade in primatology offers an excellent case study illustrating how feminist perspectives can lead to more balanced observations and to the construction of more insightful, comprehensive models.

From my first book on gender roles (*The Black-man of Zinacantan*, 1972, written when I was an undergraduate at Radcliffe) to the chapter for this volume, I have moved from relative acceptance of traditional tenets of my discipline to an iconoclastic, female-oriented perspective on biology and evolution. Like most feminists of my generation, no one taught me to think this way in college. My views developed primarily in reaction to obviously androcentric assumptions and conclusions that pervaded my field. In my book on *The Langurs of Abu: Female and Male Strategies of Reproduction* (1977), I gave equal weight to both sexes. Not surprisingly, responses to that book focused on male reproductive strategies. Partly in reaction to this, but primarily in response to erroneous stereotypes of females that were widespread in physical anthropology and sociobiology, in my next book, *The Woman that Never Evolved* (1981) I concentrated on females.

The course of my work documents a history of increasing polarization as I responded to past imbalances and distortions in the study of primate evolution. At present, I fear the pendulum may have swung too far. It is time for a new generation of anthropologists and primatologists—researchers trained in an atmosphere both more balanced and self-aware than that of Harvard in the 1970s—to attempt to synthesize the extreme viewpoints of the last two decades.

Marion Namenwirth

While growing up as an only child in New York City in the 1940s, my chief interests in life were animals, sports, science, and math, in that order. Agog at the wonderfulness of my junior high school science teacher, Miss Monks, I decided I wanted to be a biochemist. My parents and my Uncle George, himself a physician, said I should go to medical school because it was a surer way to make a living. Having recently fled Nazi Germany in mid-career, my family felt it was important that children be trained in a kind of work they could easily take with them from country to country.

I attended Cornell, City College, and the University of Minnesota during my undergraduate days; then went on to get a PhD in Developmental Genetics from Indiana University. My prospective major professor questioned me at some length about how I could reconcile a research career with marriage and family life, but this was again far outweighed by his support and enthusiasm for my research once I convinced him that my career wouldn't stop. In his lab was a woman scientist who kept complaining and grieving about sex discrimination in Academe, which she believed was keeping her from getting a faculty position (in the late 1960s). I felt she was excessively pessimistic. I knew that science departments at major universities were not hiring women faculty but, after all, they hadn't received my application yet. If I had to be better than the men applicants, that was just a more exciting challenge. (What a dummy!)

Suddenly in 1971, when I began applying for faculty positions, all those all-men science departments at major universities were falling over each other trying to track down a token woman. I had many good job offers and chose the Zoology Department at the University of Wisconsin. During my years there, they tenured every assistant professor, all of them men. In fact, the Zoology Department had never denied tenure to anyone.

When I was considered for tenure in 1977, the department debated the matter for 3 months, then came up with such a weak and vacillating recommendation for tenure that it was denied by a higher administrative committee. The reason given was that I had insufficient numbers of research publications. The man candidate considered immediately after me by the Zoology Department, who had fewer research publications than I and none at all on research done since leaving graduate school, was recommended for tenure unanimously, continuing the unbroken tenure male-stream.

I sued the University of Wisconsin under Title VII of the Civil Rights Act; the suit is now in the Federal Appeals Court. I have been doing research and occasionally teaching at the University of Minnesota while seeing my case against the University of Wisconsin through the court system.

Hilary Rose

I was born into a socialist household that reluctantly gave up its pacifism with the rise of Nazism. One of my joys during the last few years has been to see my mother, now widowed and in her 80s, find her pacifism again, her feeling of caring for the natural world and for other women. Although religion was not a formal part of this socialist background, ours was always an intensely moral socialism; this meant that if you saw an injustice and you did nothing about it you felt lousy. The consequence of this background was that I learned about individual and collective social protest very early in my adolescence. It was an interesting but uncomfortable existence, leading amongst other things to an intense hatred for the posh, fee-paying girls school where I was a "scholarship girl." Disaffected and mutinous, I neglected my studies. When I was about 17, the headmistress suggested that I become a secretary. I remember bursting out angrily, "I'm not going to be some man's brains." The nice part of the story was that my mother was supportive. Frightened by the class attitudes of the school, she nonetheless was committed to a woman's right to what she called a *proper job*, although she never would have described herself as a feminist.

My education began after I left school, like a lot of the mature students I get particular pleasure from teaching. Politically very active in the anti-nuclear movement, I was also very involved in community politics. In a personal tragedy, I was left a widow at 23, with a 3-year-old child and no very visible means of support. A year later I enrolled in the London School of Economics to study sociology and social policy. I fell in love with the world of ideas and had the incredible luck to be tutored by the last of a wonderful generation of women. They were clever, incredibly supportive academically and personally, and they were not married. When I think about them I no longer want to write about myself but about them, so that others can understand how a really practical feminism can change lives. The experience was decisive.

The expansionist years of the mid-1960s meant that a number of women were taken on as tenured university teachers, even in elite institutions such as the London School of Economics, where I taught for 10 years. My academic training and my politics left me with two passions: social policy and science. Since then, I have continued to ride these two horses, frequently falling off one or the other or both, as they have been very different in temperament and made unwilling partners. Always an active socialist, the women's liberation movement transformed my life and, slightly more slowly, my research. It was not, for example, until 1974 that I began to write about the new reproductive technology and its threat to women in this society, and more recently wrote a feminist critique of scientific epistemology, *Hand, Brain, and Heart*, which appeared in *Signs*, and also had so much pleasure contributing to this book.

My present two horses are books: one on feminism and the natural sciences, another on the social movements of poor people for change in the welfare state, which focusses on the part women have played in this.

Sue V. Rosser

I am an associate professor of Biology and coordinator of Women's Studies at Mary Baldwin College, in Staunton, Virginia. At the college, I am also coordinator of the Division of Theoretical and Natural Sciences, including the departments of biology, chemistry, computer science, education, mathematics, physics, and psychology. I very much enjoy the atmosphere at this small liberal arts women's college, which allows me to pursue interdisciplinary work in science and women's studies. However, it took me several difficult years even to begin to integrate my feminist work with my scientific pursuits. My undergraduate experiences as a woman student desiring to become a scientist caused me to seek out other women in science and, ultimately, led me to feminism. After I received my PhD in Zoology from the University of Wisconsin–Madison, and while a postdoctoral fellow, I began teaching in the University of Wisconsin's women's studies program during the first year of its existence in 1976.

Since then, I have taught courses in both biology and women's studies programs at the University of Wisconsin–Madison and Mary Baldwin College. I have written numerous publications dealing with the theoretical and applied problems of women and science. More recently, my work has focused specifically on the teaching of women's studies in science. *Teaching Science and Health from A Feminist Perspective: A Practical Guide*, published by Pergamon Press, includes much of the work I have done in this area. As a consultant for the Wellesley Center for Research on Women, I worked with faculty at several institutions that are attempting to include the new scholarship on women in the science curriculum. During 1984–1985, I was visiting lecturer in Biology and Women's Studies at Towson State University under a grant from the Fund for Improvement for Post-secondary Education (FIPSE).

Susan E. Searing

In mid-1982, I left the Reference Department at Yale to become the Women's Studies Librarian for the University of Wisconsin System of 26 campuses. My work involves bibliographic research, collection development, library skills instruction, and reference assistance to faculty, students, librarians, and university administrators. My office publishes *Feminist Periodicals: A Current Listing of Contents* and *Feminist Collectons: A Quarterly of Women's Studies Resources*. I am the author of *Introduction to Library Research in Women's Studies* (Westview Press, 1985) and am currently collaborating on an annotated core bibliography of books in women's studies published between 1979 and 1985.

In the fall of 1984, I coproduced a videotape for use in women's studies classrooms, *Women and Science: Issues and Resources*. The program features four women scientists, who share their views on science and its relations, historical and present, to women's lives – plus a segment on effectively using the library for research in this emerging field.

I'm fascinated by the challenges feminism poses to the patriarchal foundations of science and inspired by the minority of women who choose careers in science. Still, I find great personal satisfaction working in a "women's" profession. Librarianship affords me opportunities to immerse myself in the vast body of women's studies scholarship and activist writings, to experiment with feminist management within the university bureaucracy, and to build rewarding networks with feminists both within and outside the academy. In my scant spare time, I enjoy walking, cooking Indian meals, and losing myself in novels.

Mariamne H. Whatley

As a graduate student in developmental biology at Northwestern University, I became actively involved in a Women in Science group that served as both a support and consciousness-raising group for women students, staff, and faculty. When I moved on to a postdoctoral position at the University of Wisconsin–Madison, I looked for ways to continue doing feminist work in the sciences. My own area of research, the role of the bacterial cell surface in pathogenesis, did not lend itself easily to feminist reinterpretations. I decided to teach in an area that would use my biological science background in a feminist context. The Women's Studies Program had just introduced a new course entitled Women and Their Bodies in Health and Disease, and I became the lecturer while still doing research. As I became more committed to teaching women's studies, coteaching Biology and Psychology of Women and Childbirth in the United States, I found I could not resolve the contradictions between teaching and the research I was doing. I gave up laboratory science and put my energy into teaching women's health, both at the university and in the community. I currently am an assistant professor, holding a joint appointment in Women's Studies and the Department of Curriculum and Instruction, in which I direct the Health Education Program. I am particularly interested in seeing that the theoretical work on feminism and science reaches people outside women's studies. By teaching health teachers, I feel I can help implement changes in the way health is taught in the schools, both on a secondary and elementary level. My work critiques the way science and health are taught, both in terms of curricular materials and pedagogy. I also hope that my teaching, particularly in women's health, has helped make the university a more hospitable and supportive environment than that most of us encountered as undergraduates, wherever we attended college.

THE ATHENE SERIES

An International Collection of Feminist Books

General Editors: Gloria Bowles, Renate Klein and Janice Raymond

Consulting Editor: Dale Spender

MEN'S STUDIES MODIFIED The Impact of Feminism on the Academic Disciplines
Dale Spender, editor

WOMAN'S NATURE Rationalizations of Inequality
Marian Lowe and Ruth Hubbard

MACHINA EX DEA Feminist Perspectives on Technology
Joan Rothschild, editor

SCIENCE AND GENDER A Critique of Biology and Its Theories on Women
Ruth Bleier

WOMAN IN THE MUSLIM UNCONSCIOUS
Fatna A. Sabbah

MEN'S IDEAS/WOMEN'S REALITIES
Louise Michele Newman, editor

BLACK FEMINIST CRITICISM Perspectives on Black Women Writers
Barbara Christian

THE SISTER BOND A Feminist View of a Timeless Connection
Toni A. H. McNaron, editor

EDUCATING FOR PEACE A Feminist Perspective
Birgit Brock-Utne

STOPPING RAPE Successful Survival Strategies
Pauline B. Bart and Patricia H. O'Brien

TEACHING SCIENCE AND HEALTH FROM A FEMINIST PERSPECTIVE
A Practical Guide
Sue V. Rosser

FEMINIST APPROACHES TO SCIENCE
Ruth Bleier, editor

INSPIRING WOMEN Reimagining the Muse
Mary K. DeShazer

MADE TO ORDER The Myth of Reproductive and Genetic Progress
Patricia Spallone and Deborah L. Steinberg, editors

TEACHING TECHNOLOGY FROM A FEMINIST PERSPECTIVE A Practical Guide
Joan Rothschild

FEMINISM WITHIN THE SCIENCE AND HEALTH CARE PROFESSIONS
Overcoming Resistance
Sue V. Rosser, editor